AFRICAR

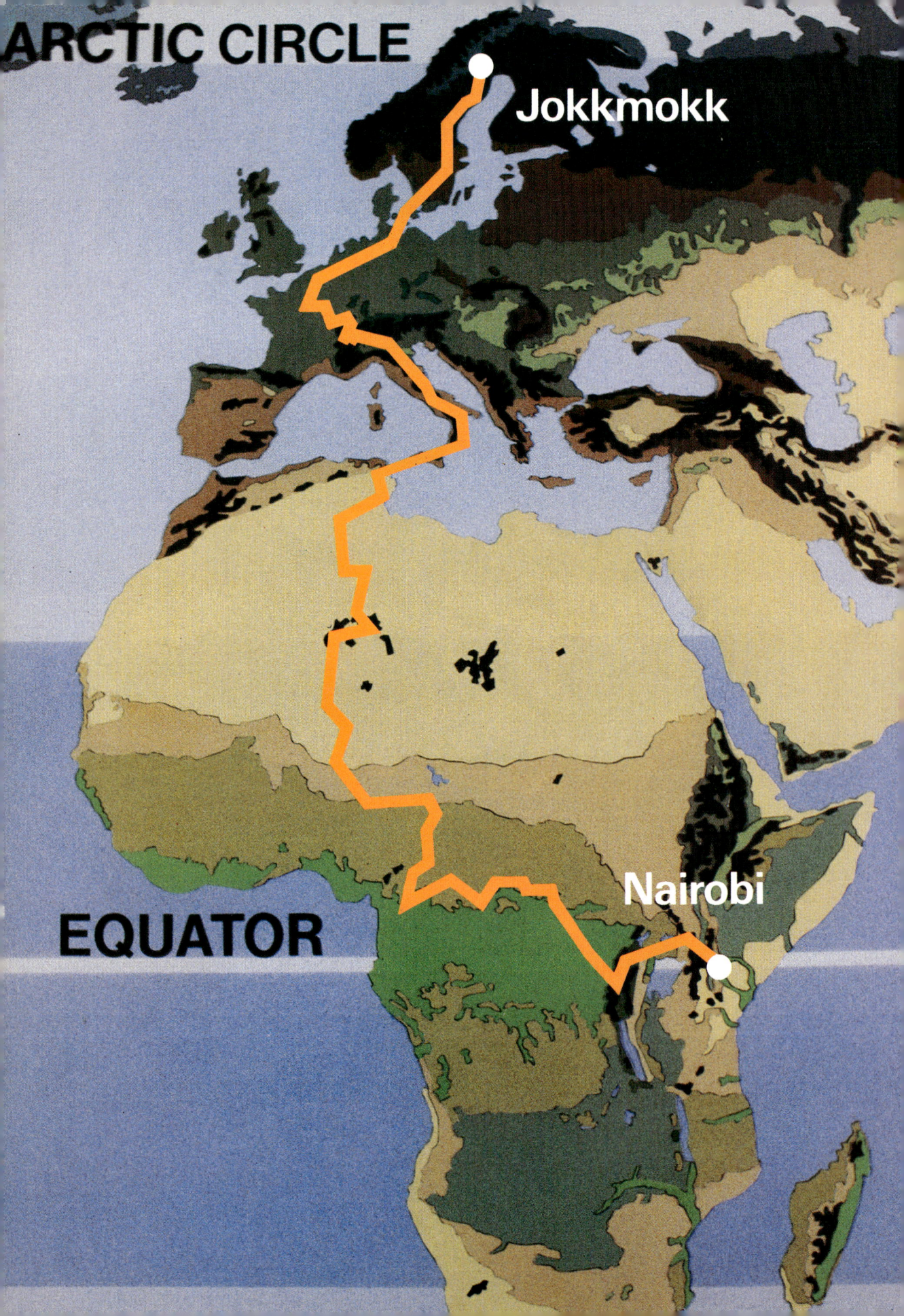

THE DEVELOPMENT OF A CAR FOR AFRICA

Anthony Howarth

Edited by Hugh Poulter
Photographs by Charles Best

THE ORDINARY ROAD

ACKNOWLEDGEMENTS

For Carolyn, who drove the wagon on her own from Tunis to Nairobi and without whose support and full participation AFRICAR would never have started. And once started, would never have reached the stage of production.

This book has evolved over a long period of time and has gone through many changes. It pre–dates my decision to actually make AFRICAR prototypes. I would like to thank those who, at the early stages, encouraged me to continue with the project. Especially Julian Friedmann, Dick Vine and Nick Webb. My thanks to Tony Hughes for the research work he did in 1983 and to my editor Hugh Poulter, also designer Dick Vine and photographer Charles Best.
Originally there was to have been a book on the history of transportation and the relevance of the motor car as just another form of transportation, not the form. It was the research for that book which reminded me of a paper I had written in 1964, entitled AFRI–CAR. Picking up the theme again, I wrote a paper in 1978 under the title AFRICAR, setting out the terms of the AFRICAR concept or system. The offer of a four–part television series from Channel 4 confirmed the decision to build prototypes. So Carol Haslam, my original commissioning editor from Channel 4 television, bears a heavy responsibility for unleashing these not–so–little funny wooden cars on the world. I am grateful for her unwavering perception that AFRICAR was a sufficiently serious concept to warrant a major television series and that it was far more than just another four–wheel drive vehicle.

Copyright: Anthony B. Howarth 1987
AFRI–CAR Copyright: Anthony B. Howarth 1964.
AFRICAR Copyright: Anthony B. Howarth 1978, 1979, 1981.
AFRICAR SYSTEM Copyright: Anthony B. Howarth 1979, 1981.

Photographs and illustrations Copyright: Godolphin Productions Ltd 1981, 1982, 1983, 1984, 1987.

Published in Great Britain by The Ordinary Road Ltd,
The Africar Centre, Lancaster

First Edition 1987
All rights reserved.

British Library Cataloguing in Publication Data

Howarth, Anthony
   Africar : the development of a car for Africa
   1. Automobiles — Africa — Design
   and construction
   I. Title   II. Poulter, Hugh
   629.2'31'096   TL240

ISBN 1–870427–00–9

Printed and bound by William Collins PLC, Glasgow

Typeset by Titus Wilson, Kendal, Cumbria

CONTENTS

1	AFRICA	1
2	AFRICAR 1963	7
3	ARCTIC	23
4	EUROPE	59
5	SAHARA	91
6	SAHEL	127
7	EQUATOR	161
8	AFRICAR 1987	195
	APPENDIX	203

1
AFRICA

You will hear more good things on the outside of a stagecoach from London to Oxford than if you were to pass a twelve-month with the under–graduates, or heads of colleges, of that famous university.

William Hazlitt 1778 – 1830

I was seven years old when the first atomic bomb was dropped in 1945. I remember the school lunches celebrating VE Day (Victory in Europe) and VJ Day (Victory in Japan) and I can still recollect the disappointment of tasting my first banana, a black shrivelled object, fed to us to symbolise the return of peace, prosperity and world trade.

Private car production simply stopped from 1939 to 1945; even in the United States it was reduced to a few thousand units after 1942. The vast majority of existing cars were greased, put on blocks to preserve the tyres and kept like the collectors' items they have turned out to be. Motoring was for essential purposes only and for a privileged few, but even they had petrol restricted to a minimum number of miles.

But life went on. Industrial and agricultural productivity rose to unanticipated heights and the distribution of essentials was efficiently and effectively organised. People went to work as well as to war, filling their leisure time without the aid of television. To counteract the ghastly waste of war, civilian society enjoyed a period of almost unparalleled conservation and efficiency.

In a time of crisis Europe fell back on its large public transport infrastructure and in small countries with short average journey distances, people started walking again.

The continent of Africa is vast. Second in size only to Asia, Africa is almost twice as large as South America. The Sahara desert alone could swallow the whole of the continental United States – twice. The Republic of Chad is two and-a-half times the size of France. Zaire is ten times the area of its old colonial overlord, Belgium – and likewise, Kenya is more than twice as big as the United Kingdom.

Yet Africa has few paved roads, even fewer convenient navigable rivers, hardly any railways and little public transport.

After one hundred years of the internal combustion engined automobile, the list of vehicle makes and models that can claim to have been in any way suitable for use in Africa, is short and select.

Most of the world's cars and light commercial vehicles are built as disposable consumer products for the tarmac–borne industrial nations. Even in the 1980's the motor car is a luxury used by less than a sixth of the earth's population.

It is in Africa that the car and the light commercial vehicle are essentials rather than luxuries. One ton pick–ups and station wagons provide for the movement of people and goods in remote as well as populated areas.

There was one vehicle that surpassed all others in usefulness throughout Africa. It was French, it was not four–wheel drive, it came as a car, a pick–up and a station wagon, it was apparently nothing special.

The 403 Peugeot and its descendant the

A Tunisian scrapyard ▷▷

Mountains of the Tassili region near Djanet ▷

The Kembe Waterfall in the Central African Republic near Bangassou
▽

2

△
The ubiquitous Peugeot 404 pick-up

404, four times winner of the brutally tough Safari Rally, dominated vehicle sales throughout Africa from the late 1950s to the beginning of the 1980s.

In the early 1920s, when the roads of America were little different from the roads of the Third World today, half of all motor vehicles were Model T Fords. Fifty years later almost half of all light motor vehicles in Africa were Peugeots.

Now the Peugeot 404 is out of production. For Africa the loss of a suitable vehicle can, in the long run, be as devastating as a drought.

But even the Peugeot was never quite a car for Africa. It was limited to the roads, the better tracks and the firm desert pistes. The Peugeots, although sometimes locally assembled, were not practical for actual manufacture in Africa.

But then conventional industrial wisdom teaches that local vehicle manufacture in appropriate numbers for a typical Third World market is unthinkable.

In this case conventional wisdom may be wrong.

2
AFRICAR 1963

I took the photographs in this chapter in Southern Africa in 1963

Most people living in Western society understand ordinary roads to be tar–sealed highways with defined curbs, centred white lines and clear directional signs.

This is far from the truth. In reality, more than eighty percent of the world's ordinary roads are loose surfaced, rough, potholed and rutted.

The motor industry and its allied trades occupy about a third of global industrial activity. Yet two percent of all vehicles are in the 'Third World', where almost three quarters of the planet's population is to be found.

As a result of this, almost all modern motor cars, commercial and four–wheel drive vehicles are completely unsuited for use on the ordinary roads.

But the problems of potholes, corrugations, mud, sand and visibility, are just obstacles that can be overcome by applying the basic laws of physics, a little ingenuity and some common sense.

There are other, more important problems to be encountered along the way.

It was dark when they took me to the small wooden hut. I had no idea where I was. It could have been in a town or in the bush. Anonymous and bureaucratic, the hut was typical of official buildings in the area. Tongued and grooved boards, cross braced on the inside once painted with white gloss enamel, now a chipped and yellowed cream. There may have been two or even three rooms – I only saw one.

The wooden chair was hard, facing a desk, bare apart from a fat, well thumbed buff file, a ragged bible and a lamp. The lamp, bronze coloured with a flexible stalk and a half–cylinder metal shade turned to shine partially into my eyes, shadowing the hostile figure examining me from the other side of the desk.

The set and circumstances were straight out of the 1950s B movies. Saturday matinee at the local town cinema – the good guys versus the bad guys and not much to choose between them.

A huge hand with short thick fingers crashed on the desk.

'Where is your wife?'

Afrikaans can sound like Hollywood World War II German. But this was no movie, it was real and I was frightened. The date, early December 1963; location, somewhere near Mafeking and the file on the desk was mine, the security car that had rushed it from Pretoria was parked outside, its engine still warm.

Where was my wife? A good question to which I had no sensible answer. I thought she may have gone to Cape Town but she could be in Swaziland or even on a Transkei mission station – I actually had no idea.

'I think she has gone to Cape Town; she is taking the ship back to England while I am...'

'You think?'

The gross hand crashed on the desk again, punctuating the sarcasm.

'You came here six months ago with one wife, name; Tanja. A good looking blonde if this photo is not lying. Now you say you have lost her!'

'I know she is going to Cape Town to catch the ship. I have to...'

Crash! This time the whole room shook

and a revolver, hanging complete with a cartridge belt from a rusty nail on the wall, started to swing gently – ominously.

'She is going? What we want to know is where she is now, what she is doing and why she is not with you?'

I had been trying to explain that Tanja was going back to England by ship while I drove our borrowed car 4,000 kilometres back to Nairobi. As to where she was and what she was doing, who could tell?

'I don't know.' I replied firmly and honestly, it was, after all, the truth but evidently it didn't come across that way.

The desk, the floor and the walls shook again. It was pointless as the man was not interested in anything as banal as travel arrangements. He knew what he wanted and I knew I couldn't give it to him.

The file on the desk was testament to more than ten previous interrogations in spartan rooms of grubby police stations in every province of South Africa. Johannesburg, Cape Town, Port Elizabeth, East London and Durban as well as several small dorps, unmemorable and best forgotten. Now for the first time, these reports were all together in one incriminating folder.

It was the file that shocked me most when I was pushed into the hut and uncuffed. Previously it had been a six month game of hide–and–seek, of slipping over convenient borders to release the tension, of fighting an ulcer conceived of greasy food and cheap local brandy – but suddenly it was no longer a game. It had become official, formalised and recorded into the bureaucratic machine of the state security department.

It had taken them six months to put it all together and they had just managed it in time to catch up with me as I tried to slip across the northern border into Bechuanaland and the relative freedom of black Africa, for the last time on this trip.

The man eventually grew bored with the questions about my wife. He had brought tea for both of us, offered me a cigarette and placed the revolver on the desk in front of him. All on cue, all in the script. Predictably he started a new attack.

'Where are your photos?'

Suddenly he looked more menacing. Everything from his 120 kilogram girth to his polished Yul Brynner scalp exuded hostility.

'We...'

The homely Royal 'We' had changed into the oppressive State Machine 'We', as in: 'We have ways of making you talk'.

'We want your photos, all of them.'

He started to play with the gun.

'I haven't got any photographs but there are 150 rolls of film on the back seat...'

This time the room exploded in a wave of pain. The butt of the revolver would have come down on the desk had my left hand not been in the way. At least three knuckles seemed to be crushed.

'We don't give damn about your film, we want your photos.'

Through the pain I recognised my dilemma. All my processed photographs had left the country, many of them taped around the waists and concealed under the cassocks of priests going on furlough to the UK. All I could offer was exposed but unprocessed 35mm black and white film. But film was not in the interrogator's brief, only 'photos' would do.

I have never been particularly brave, especially in the anticipation of pain or danger, but I do have a temper that under attack, shows itself as an icy anger. A year earlier in Pakistan I had a knife pushed into my throat while two other would–be muggers held my arms. In sheer cold anger I spat into the knife–holder's face causing instant disgust, offence and recoil, giving me the opportunity to escape.

Now, despite the pain and intimidation, I became furious with this bullet–headed boorish Boer. He was an idiot. One hundred and fifty rolls of film, left openly on the back seat of the car only needed a few hours laboratory work to turn them into the 'photos' he wanted.

I don't know how much time passed, it was always the same question followed by the same answer. There was no more overt violence, the blow to my hand could have been a mistake born from years of table thumping but it could have been deliberate – there were plenty of threats.

'We reeely hate people like you, we can put you away for years and your fancy British passport won't help you here. We hang people here for treason and much less.'

'We want your photos.'

'I have no photos, only 150 rolls of film on the back seat of my car.'

The interrogation ended as suddenly as it had begun but not before an ironic scene had been played out, a scene that even now, more than twenty years later, seems as vivid and as incredible as it did then.

The man took a list out of his desk drawer and asked me to read the names on it carefully and tell him if I knew any of them. It read like a Who's Who of South African anti–government activists. I recognised many of the names but did not personally know them.

'No, I don't know any of these people.'
'Good.'

He looked pleased in a fatherly sort of way.

'Now we will make sure you don't get to know them on your way through Bechuanaland.'

He took my undamaged right hand and placed it on the ragged Bible.

'You are a Christian?' he asked.

Unwilling to go into the vagaries of beliefs and because I had been working in South Africa as a photographer for a large missionary society, I took the easy way out.

'Yes,' I said, perhaps a little hesitantly.

'Good, then you will swear on this Bible that you will not make contact with any of these people while you are in Bechuanaland. You will swear to each one individually by name and then I will let you go. Don't ever come back.'

It was not easy to look my inquisitor in the eye. Christian or not, I sense that old things have a presence and a power and the Bible is a very old book. As I felt the warm leather under the palm of my hand, childhood taboos pulled gently at my conscience. I was wondering whether it was an English or Afrikaans Bible when the full irony struck me.

Many, if not most of the people on the list would, by Western standards, be devout Christians. My inquisitor belonged to the same church, followers of a tradition that Luther or Calvin could hardly have predicted; the perverted protestant heresy of white superiority maintained by violence in the name of the Lord.

The taboos drifted away. I wanted to laugh out loud as I solemnly swore that I would not talk to or make any kind of contact with over fifty well known political refugees. Several of the names were of people I wanted to meet, at least now I knew which country they were in.

It was late afternoon when they dropped me back at the border. My car was there, apparently untouched. They handed me my passport, made it quite clear that I would not be welcome in South Africa again and opened the five bar gate that stood between me and the north.

I started the car after establishing that I would have to change gear with my right hand, drove through the gateway and on to the deep rutted track that, after a thousand kilometres, would lead to Francistown and the White House.

Ten kilometres past the border I stopped. On the back seat of the car there were 150 rolls of unprocessed film. They represented my last month's work, more than 5,000 images from Soweto, gold mine compounds, Kimberley and other places and people along the way.

The assignment, to record South African society with emphasis on the effects of the Group Areas Act on all races, was complete.

As I drove, nursing my left hand, I recalled the words of the Minister of Information at a bizarre meeting in Pretoria some six months earlier. My employers, good Christians that they are, had sought ministerial approval for the assignment. This had been granted verbally with a great deal of double talk about obeying the law. Of course such obedience would have made the assignment impossible as whites were not allowed into black areas without a permit as much as blacks were not permitted into white areas without a pass book.

I had ended the interview by asking for a short letter confirming the ministry knew about me and my proposed itinerary, in case I was stopped by the police.

'No need for that, Mr Howarth. This is not a police state,' was the Minister's reply.

I stopped at the first town, a few shacks and corrugated iron roofed stores stood back from the road in the sand and the scrub. A simple bar provided beer, food and conversation. After months in South Africa I found it hard to believe that I could wander into the first bar I came across, find it full of African people and just merge into the background.

Even then it was difficult to resist watching the door, alert and listening for the sound of screeching brakes, the crunch of tyres and slamming car doors, the tell–tale sounds of a police raid, south of the border.

The bar was mostly full of refugees, drinking their evening away with 'European booze' (bottled beer) then illegal to blacks in South Africa. They knew about me as word had spread that I had been stopped and taken away by security at the border. They were inquisitive to know who I was and what I had done. We all laughed, almost hysterically, about the story of the list and the Bible. They gave me directions to the White House in Francistown, showing once again the degree of naive trust that has been the undoing of so many South African anti–government resistance movements.

An His Master's Voice wind–up gramophone played scratchy Qwela, Masikele's haunting trumpet and the close harmony of the Dark City Sisters, a tragic blend of European influences combined with traditional Zulu and Xhosa songs. A sound once heard and understood, never, ever forgotten.

Someone put on a well worn record of Nkosi Sicelele i Africa, God Bless Africa, then almost a national anthem for half a continent, now a song symbolising freedom for the black people of South Africa. The emotional effect, as always, was the equivalent of the playing of the Marseillaise in the film Casablanca.

I found myself dancing or oscillating, as the Pata Pata can only be described, with an ochre girl barely half my height. She had that unmistakable part Hottentot, or Bushwoman figure and the sad, soft face of Southern Africa. Her name was Tande.

She took me to bed in a bare room at the back of the bar, washed me, oiled my injured hand then took away the tensions of months. Nothing asked, everything given.

The bed was too short and had no linen, just a stained and curled mattress on an iron frame. The bed head was against a low window, so dirty that curtains would have been irrelevant. Nothing mattered, it was a long time before we slept.

When I awoke she had gone.

Today the north to south road on the east side of Botswana is a good tar sealed highway joining the frontier with Zimbabwe to the southern border with South Africa. In 1984 I travelled this road as a passenger in a Mercedes Benz with the cruise control set at 140 kilometres per hour. The air conditioning isolated the driver and myself from the hot, hostile environment at the edge of the Kalahari desert. We had no punctures.

In December 1963 there was just a single, rutted track leading along the edge of the desert. The ruts, deep and confined to the track, were unlike the Sahara desert where they spread out over hundreds of metres or even kilometres. The fierce thorns guard the track like sentries, with a base over ten millimetres thick they are capable of piercing any car or truck tyre. Drivers have to stick to the ruts keeping clear of the scrubby thorn trees either side in the hope of avoiding punctures.

The drive to Francistown took three days. The small 1100cc car I was using proved difficult to manage as the wheels tended to fall into the ruts causing the sump guard to plane along the raised sand centre until the wheels were virtually off the ground and the car would literally grind to a halt.

The alternative was riding with one wheel deep in a rut and the other up on the edge or the centre. Riding the edge, rather than the centre, was easier and safer but resulted in frequent punctures. Riding the centre was dangerous as the soft sand could drag the car so that it was suddenly out of one rut and into the other, an attitude change of 45 degrees, enough to make a car turn over at the right speed.

The car suffered fuel starvation problems due to the extreme driving angle. Continuous fuel pick up was sometimes impossible to guarantee even with the tank half full. Punctures were a nightmare and virtually impossible to repair with one useless hand. I had to siphon fuel into the carburettor to get moving again and had great difficulty unsticking the car on my own when it grounded itself on the centre hump.

There are a number of techniques open to the lone driver but the simplest goes something like this: Set the slow running to between 1200 and 1500 rpm (a bit faster than usual), leave it running in bottom gear with the front wheels turning, then rock and push until the car starts to slither forward. Keep pushing until better ground is reached then run alongside and jump in without tripping and falling. I have not lost a car yet this way but one day I am sure it will happen.

CARELESS DRIVER SOUGHT
Nationwide Search

MP asks, 'Are cars safer without drivers?' Yesterday a driverless car was seen to pass through several villages in the Southern region. The car proceeded slowly and cautiously, well within the local speed limits. The police did not intervene.

A spokesman said: 'The car was driving very carefully and not breaking any traffic laws. Off the record that's more than can be said for most drivers around here on a Saturday night.' The car was last seen heading north–west towards the frontier, where, at current rates of duty, its value will treble the moment it passes the barrier. MP The Hon...

The most innocuous of the many fantasies of the long distance solo driver.

Many people would find it hard to believe that the small family saloon I was driving now coped with these conditions better than a Land Rover that I had driven over the same route some three years earlier. The Land Rover also did not have sufficiently wide track to fit the ruts and the ground clearance, at about 8.5 inches under

the differentials, was also not enough. The huge axles would plough into the raised centres like anchors, lifting the wheels off the road.

The smooth underside of the saloon car, combined with all independent suspension, allowed the wheels to go down, beyond the 8.5 inches of the Land Rover's solid axle, providing enough grip and support to get me out of trouble.

It didn't take an engineering degree or a particularly perceptive driver to realise what was needed on this and many other types of 'ordinary roads'. A lightweight, wide tracked vehicle with a clearance close to 12 inches, a smooth sledge–like underside and independent suspension on all four wheels. A vehicle of these specifications would only need four–wheel drive to handle extremes of mud, snow or soft sand on steep gradients or under true off–road conditions.

At this time those guidelines were to a certain extent recognised in Southern Africa as the most popular rural vehicles were not chosen from the limited range of four–wheel drives, but from huge, wide tracked and relatively large wheeled American pick–ups, notably Chevrolet. Although heavy, thirsty and atrocious on bad roads, with some load in the back to provide traction for the rear driving wheels, they managed pretty well.

In the mid–1980s Gaberone, the capital of Botswana, sports a population of about 200,000, has several luxury hotels, motels and casinos. There is a massive, booming industrial complex capable of spending close on $1–million a month on computer and office equipment.

In 1963 I almost missed Gaberone and had to turn back to find it. There were white pegs in the sand to mark possible future city blocks and just a couple of administrative buildings with some temporary accommodation.

I had a contact in the government and was invited to attend a wedding the next day.

It was a huge and surprising event, everybody out in their finery. The bride's white was startling in the harsh sun next to her dark skin. This was a chief's wedding and the new Prime Minister Seretse Khama was there with his European wife. People were polite, lively and gracious, sipping champagne and nibbling wedding cake while talking of London and Paris as if they were just down the road. They did not mention the nearest big city – Johannesburg.

Fresh from South Africa I found the wedding a symbol of hope. The fact that the newly elected Prime Minister, soon to be designated the first President of independent Botswana, was black and openly consorting with his white wife when only 500 kilometres down the road they would be beyond the law, was both enlightening and rewarding.

Over the southern border he would probably be charged with rape and would be lucky to get out of prison alive while she would be cast as a 'kaffir–loving whore' and, if she escaped prison, would be ostracised by her peers. What I was witnessing just couldn't happen there yet here it was happening with dignity and normalcy.

It is easy to dismiss an event of this kind – the new African elite taking on the mantle of their former colonial masters and playing, while around them most of the population exists in extreme poverty. In some countries this has been the case, but here it seemed different. Bechuanaland has never been a colony in the normal sense but has been a British protectorate since the end of the First World War, initially under the League of Nations and lately the United Nations mandate.

The new government, and without question the ruling class, were mostly born to traditional positions of authority within an established social system. During the first few years of independence Seretse Khama achieved a great deal, setting Botswana on a course that has allowed the country to develop within as stable a network as any I know. He started walking a fine line between independence and South African influence and achieved clear independence despite shared borders, monetary and transport systems.

In 1984 I found the floating Botswana currency, the Pula, meaning 'let there be rain', was standing a good twenty percent higher than the South African Rand. The outside advisors and civil servants, not drawn from South Africa, were highly professional administrators from Holland, Britain, Scandinavia and other European countries. Each one was there to do a professional job on a contract basis, not there to acquire land and involve themselves in local business.

The following day I was heading north again, the road becoming stony. It was easier and faster, but the noise of the stones thrown up by the front wheels was alarming as if they would come straight through the bulkhead and wheel arches into the car. It was a particularly loud report which sounded like a shot being fired that reminded me of the gun.

In the rush to get out of South Africa I had forgotten my gun. I hadn't declared it at the Bechuanaland border and didn't even know if it was still in the car. Try as I may I couldn't get the glove compartment to open with my damaged hand so I had to stop the car to check it. The gun was still there, wrapped in an oily rag, beside it a box of 180 rounds of ammunition. Twenty had been fired at a rock in a Transkei river as practice, few had hit it.

This was the first and last time I will possess a hand gun. I can't actually recall why I bought it in the first place. It had a lot to do with a more innocent past, growing up on a farm where guns were used daily to crop rabbits, provide food for the dogs and pigeons for the pot, shooting was a part of life.

A hand gun is something else, it is useless for shooting rabbits or pigeons. It has a single purpose which is to kill human beings at close range. A South African born Life Magazine photographer, Grey Villet, had argued convincingly that he would not go out at night in Johannesburg without a gun because of white muggers.

I bought the gun in London's Piccadilly. Straight into Cogswell and Harrisons'.

'I would like to buy a gun.'

'And what sort of gun would Sir have in mind.'

The assistant turned to indicate a rack of sporting rifles and shotguns.

'Something like that,' I said indicating a small revolver under the glass topped display counter.

'And where is Sir going?' He asked me as he reached under the counter for the offending item.

'South Africa.'

Apparently this was a satisfactory answer, the assistant nodded knowingly and broke open the chamber.

'This is actually a lady's gun, only five shots. The cylinder is slim and the barrel short to slip into a handbag, nice and light.'

He offered me the gun butt first, it was light and very neat.

There is something seductive about guns. The precision feel and fit, the high quality oiled steels and ergonomic hand grips. It struck me as being strange that killing machines are so well designed and made in a world where most household items are manufactured without a thought for the unfortunate end user. Rifles and shotguns are frequently tailored for their owners, but how many car seats and steering wheel positions get the same individual attention.

'It is a .38,' the assistant pointed out, evidently concerned that I had been reading my James Bond and doubting its stopping power.

'Of course it only has a short chamber and won't accept the high velocity cartridge, but it should hold anybody at 10 yards.'

I told him I would take it and 200 rounds of ammunition. I gave him my flight number and asked him to have it at the airport for collection.

Customs gave me a receipt: One Smith and Wesson .38 revolver and 200 rounds of ammunition. The box was passed directly on to the purser of the aircraft for collection in Cairo.

Oddly enough, the gun became my passport for getting through borders without the usual batch of questions regarding my five Nikons, 17 lenses and 400 rolls of duty free film.

At Cairo airport I was taken without ceremony to a dusty office, littered with papers and files and inhabited by an Egyptian Chief of Customs, a heavy man in his mid-fifties, who looked me up and down with the wisdom of 10,000 years of civilization.

'What do you need a gun here for? You going to shoot the president?'

'I don't need it here, I'm going to South Africa.'

His eyes narrowed as I mentioned the

unmentionable country.

'You South African,' he asked paging through my British passport.

'No, English.'

'In which case you need it there, not here.' He wrote out a receipt and the box, unopened, was placed in his desk drawer.

At Nairobi airport there was peaked cap, polished shoe, European efficiency. The box was confiscated and I was instructed to report to the firearms bureau the following morning.

I duly appeared at the wooden, pre-fabricated hut to find the officer in charge in regulation shorts, open necked shirt, knee length stockings and a peaked hat hanging on the back of the door. He resembled Trevor Howard in a safari role.

'Why have you come to Kenya?'

'To take photographs of the mission stations for the Society of the Propagation of the Gospel.'

He smiled – 'With a gun?'

'And to cover pre-independence elections,' I added as if it explained everything.

'The emergency officially ended three years ago,' he paused, then looking directly at me, said:

'But the firearms regulations have not changed. I can give you a license but if you lose the gun the maximum penalty for you is death by hanging. I suggest you let us look after it while you are here.'

Tanganyika customs, still all European despite independence only half a year away, just added their own seal. The box had still not been opened since it was packed in Piccadilly, but the gun had become my passport to cross borders with cameras, films and even a borrowed car without duty being charged or documentation exchanged. It was too good to last.

Northern Rhodesia was not interested in the gun, the cameras, the car or any of the letters of assignment that I carried from both the SPG and the Church Missionary Society.

The punctilious Scots immigration officer was simply convinced that I and my wife planned to become destitute and would have to be repatriated to the UK at, it appeared, his own personal expense.

'But we have a car, we can drive back.'

'How can you be sure the Tanganyikans will let you through?'

'We will go and ask them now.'

'Whatever they say now may not be true later.'

'We are working for the SPG and the CMS, they will cover all expenses.'

'What proof do you have?'

'These letters asking the Bishop to help us.'

'Those are only letters, they are not proof.'

I began to wonder if I had sprouted long hair and a beard. I felt my face and my head. No, clean shaven and clean cut.

'We can sell the cameras.'

'Not without paying duty.'

The European bureaucratic mind is possibly the most immovable of all. The French are certainly top of the list with the Germans coming second, but the English are a close third. It is not a competition to be proud of. Travellers today who constantly complain of the 'Third World' bureaucrats, would do well to remember who taught the system.

They may find it was themselves.

The immigration officer grudgingly agreed to send a morse coded message from the signal room that operated only between 4 and 4.30 pm each day. The message would ask confirmation of our status from the Bishop of Southern Rhodesia, into whose care we had been placed. He then officiously made us both prohibited immigrants of the Federation of Rhodesia and Nyasaland, before he sent us back to Tanganyika to await the reply.

Now in no-man's land, we were offered the local cell, a thatched mud hut, as accommodation, rather than officially re-entering the country – it would save a lot of paperwork to treat us like criminals.

Two days later the reply came back. It was curt and to the point. The Bishop of Salisbury had never heard of us. There was a gleam in the immigration officer's eye. It took us a while to realize what had gone wrong and a further two days to persuade our inquisitor to send another message. He had addressed the first message to the Bishop of Salisbury, who indeed had never heard of us, it should have gone to the Bishop of Bulawayo.

Five days after reaching the border we were on our way again, but this time with uncancelled full page prohibited immigrant stamps to the Federation of Rhodesia and Nyasaland in our passports. So what, in six months time there would be no Federation. Instead, out of its unlamented ashes, there would be Zambia, Malawi and Rhodesia. The stamp would have little relevance, except perhaps, in Rhodesia.

There was no border between Northern and Southern Rhodesia, just a curious barrier of white disliking, even hating, white, as they watched each other from either side of the Zambesi.

At the Otto Beit Bridge into South Africa, they were only interested in the cameras, not to levy duty, but because of their propaganda potential. The gun was no problem, a license was obtained immediately and the box broken open to confirm the serial number. Somewhat to my surprise, there was a short barrelled five shot Smith and Wesson .38 revolver with 200 rounds of ammunition.

Now, standing beside the main North–South road in gentle Bechuanaland, I realised for the first time I was illegally in possession of the gun and God only knew what penalty that carried.

Beyond the simple, if potentially fatal, legalities, something else had happened when I was in South Africa.

It had been in the lounge of a white priest's house in an affluent Johannesburg suburb. We had been talking about the horrors of apartheid, in what seemed to me, a rather unconvincing way. The priest looked after a wealthy, all white parish and I doubted if any member of his congregation would stay had he introduced a single African into church on a Sunday morning.

The priest's son, who had a heavy Afrikaans accent, was obviously having trouble with our conversation. Later I was chatting with him when he started to talk about guns.

'Oh, I've got a gun,' I said,'It's upstairs, I'll go and get it.'

I brought the gun down as a gesture of friendship to include into the evening someone of my own age who was obviously embarrassed by our endless political discussion. It had also crossed my mind that if we went too far he might be the sort to inform the police. It had been known to happen in the best families in South Africa.

It is a perverted society that can make one feel secure while showing off a gun in public, but insecure while discussing political issues.

As I took the gun out of the box I knew the priest was staring at me. When he spoke there was none of the previous warmth in his voice.

'We don't allow guns of any kind in this house, I don't know what they mean where you come from, but here they symbolise something which apparently, for all your talk, you do not understand.'

With that he left the room.

I did understand, but only as he said it. Unable to find a hole to crawl into, I packed the gun away and took it outside to the car. I don't recall that the priest ever spoke to me again.

In Francistown I quickly found the White House. It was in the African location, which in Bechuanaland simply meant the poor end of town – none of the barbed wire and police posts of South African locations.

The house was low and more humble than I had anticipated. There were, after all, more than fifty South African refugees holed up there, stuck, without money or transport, tolerated by the Bechuanaland authorities who would have preferred them to go away. They would have done had the plane they chartered earlier in the year to Dar es Salaam not been blown up on the Francistown airfield, reputably by South African agents.

These were the heady days of Pan Africanism – the great dream of Kwame Nkrumah. The PAC or Pan African Congress, was the militant black movement in South Africa. The older, and well established ANC, or African National Congress, was taking a moderate non–violent stance, giving white South Africans, had they the genuine goodwill and intelligence required, possibly the last chance to form an alliance with the blacks. Today the militancy of the modern ANC makes the PAC of the early Sixties look tame by comparison. Then it was a skirmish – now it's war.

People, mostly young men, were sitting or standing around outside the house. The mood was decidedly unfriendly. I was stopped well before reaching the house and asked my business. It took some explaining. I was English, driving up to Nairobi within a couple of days and prepared to give a lift to as many as would fit in the car. The maximum practical number was three if we were not to attract unwelcome attention travelling through the Federation.

In time, an older man came out and spoke with me. His name was one I recognised from the list. He apologised for the air of distrust and repeated the story about the chartered aircraft. He explained that the others thought I was a South African agent but said he could tell from the way I spoke 'with him' rather than 'at him' that I was not a South African and had never lived there. I took this as a compliment and thanked him.

We talked for a while about the state of things in the South. I brought him news of people still active and at liberty and of a white woman's arrest he was not aware of. He asked me to leave while they had a meeting to discuss my proposition. I was to return in the evening.

I fueled, watered and oiled the car, had a snack and bought some provisions for the journey. At six o'clock, in the African dusk, I returned to the White House. The air was thick with woodsmoke as meagre dinners were cooking over fires in the hundreds of shacks that made up the town.

The man came out to meet me and introduced me to my two travelling companions. One, Luke, a short emaciated man of about thirty, looked and was Johannesburg location born and bred. He wore a dirty white shirt outside a pair of equally dirty grey flannel trousers which were at least four inches too short in the leg. A pair of once white plimsoles with holes for his toes in the left one, completed the picture.

He carried a newspaper–wrapped bundle containing a shabby blanket to sleep in and a small notebook filled with addresses and contacts. At this stage I was told he was going to the Copper Belt in Northern Rhodesia.

The other passenger, who called himself Jan, was a total contrast. A huge bullet–headed Arab with a polished bald scalp. He wore fashionable trousers, a Hawaiian shirt, had socks and polished shoes and carried a brown suitcase. Jan was going to Dar es Salaam.

The plan was simple. We would leave immediately, my two passengers in the back posing as hitch hikers. We would stop about five kilometres before the border and while I slept in the car, they would walk around the border under the cover of darkness. Passing through the Southern Rhodesian border at dawn, I would collect them at a large bayobab tree set back from the road on the left, some three kilometres into the country.

It all went smoothly and we were safely on our way by 6.30 am. After some discussion, we chose to take the route through Bulawayo, despite a recent bomb incident in a supermarket. There would probably be just as much police activity on the longer route through Salisbury. Anyway, a white man driving a Kenyan registered car with a couple of coloured 'boys' in the back should attract no special attention in Rhodesia.

With this in mind, we agreed that we should be careful not to show any signs of familiarity in the presence of other people, and that, should we be stopped, they would claim to be hitch hikers. I suppose it was reasonable for me to protect myself but I knew that they both had no papers and would most likely end up in prison before being returned to South Africa. At the start of the journey I fortunately didn't know just how serious that would be for both of them.

I had set an ambitious target for the first day. Lusaka, the capital of Northern Rhodesia by 8 pm that evening. Fourteen and a half hours to cover about 700 miles or nearly 1,200 kilometres, but the roads were good and the car going well, cruising between 100 and 130 kph. It was feeling the extra weight on the hills and the little engine was affected by altitude.

Most of Europe is at sea level and mountains are accepted as obstacles. European car makers seem to have no idea that most of Africa and South America is high, in fact from Nairobi to Johannesburg you rarely drop below 3,500ft and often you are driving at 5,000ft. Even by compensating the oxygen loss by changing the carburettor jets, internal combustion engines lose power. A car like a Peugeot 404, which could cruise happily at 140kph all day in Europe, often had trouble maintaining 110kph in central Kenya. In Africa, a 2 litre Land Rover was reduced to a crawling pace on moderate inclines.

Using a press–on technique that I had evolved for long haul journeys across Europe and Africa, we made no stops apart from fuel every 300 kilometres and maintained our average. All of us were tense as we counted off the Southern Rhodesian miles. We did not know if there would be border posts at the Zambesi border.

The break up of the Federation and independence for Northern Rhodesia as the new nation of Zambia was now only weeks away so new border posts could have been set up. Beyond that possibility there was the harsh, sterile atmosphere of Southern Rhodesia, a country that exuded hostile vibrations. The law of the Federation was undoubtably more liberal than the law of South Africa, but white Rhodesians always seemed to compensate for this by being tighter lipped, tighter arsed and openly more bitter than even the Afrikaners to the south.

I should be careful. Personal attitudes and behaviour are one thing, the law is another. It is the enshrining of racial bigotry in the basic law of the land that makes South Africa so evil – far worse than countries like the Soviet Union where oppression may be equally practiced, but where the law is actually one of equality.

There were no border formalities at the Zambesi. Over the bridge we were in 'almost Zambia'. The oppression of the south lifted and we began to talk.

Jan, over a period of several months, had blown up about 70 electrical pylons on the main Reef to Durban power line. Although he went into graphic descriptions about where to put the plastique, I never could quite fully grasp the details. I had the impression he was from Cairo but he said he

was from Dar. He could have been South African, it didn't seem wise to delve too far. He had finally been identified and had to run for his life. Not surprisingly, sabotage carried the death penalty.

Luke was from a PAC cell in Johannesburg and he, poor man, was designated the Pan African organiser for Katanga. He was heading through the Copper Belt to Elizabethville at a time when anyone with any sense would have been travelling fast in the opposite direction.

The Congo and its province of Katanga were in flames and it was getting worse. The Anglo–US–Belgium mining companies were fuelling political differences in the hope of getting Katanga to permanently secede as a super–rich mineral state, thus leaving the ex–Belgium Congo to die an economic death as Zaire.

The UN, with Irish and Ethiopian troops, a deadly combination, would try and ultimately succeed in preventing this from happening. The UN troops were firing shots in anger for perhaps the last time. When they finally won the battle, the super–powers clipped the wings of the UN, no doubt so that it would never get in their way again.

Into this bloody pot was thrown the joker mercenaries, like Mike Hoare and his not-so-merry band of brigands, who eventually ran amok. Even those who had hired them were so horrified that they either abandoned them or actively fought against them. I didn't envy Luke his posting.

'Do you want a gun,' I asked, 'I think you might need it.'

'You have a gun?'

'Yes, a .38 revolver, it's in the glove pocket, have a look.'

Luke was now in the front as we felt no danger in 'almost Zambia'. Jan had spread himself, all six foot three inches of his bulk covered the back seat. How they survived the first ten hours with both of them in the back, I just couldn't imagine, The 1100 was hardly a family car and certainly not an African family car.

Luke took out the revolver, handling it delicately.

'You have ammunition?'

'Yes, nearly 200 rounds, it should be there as well.'

Years later, after being exposed to the truly violent American society as opposed to the basically pacific African society, I looked back on this incident as a perfect recipe for quick robbery and car–jacking. But innocence sometimes has its own reward. I would have no more expected my hospitality to be abused than my two 'terrorist' companions would have considered abusing it. They were not terrorists then, strictly freedom fighters and legitimate political activists. One day they will be heroes.

'I haven't got any money,' Luke said suddenly.

'What?' I asked, 'Oh, you mean for the gun.' The thought had not occurred to me, I just wanted to get rid of it and this was the perfect opportunity.

'That's okay, you can have it, I don't need it any more.' Somehow 'need' was a better word than 'want'.

We reached Lusaka at 9.30 in the evening and tried to get rooms at three cheap hotels, all of which threw my companions out for being black and me out for being with them. This was, after all, two weeks before independence.

Then I had a brainwave, let's go to the most expensive hotel in town, two weeks before independence they wouldn't dare throw out an African. And so it was. Despite our comical appearance we immediately obtained a large and luxurious room with three beds. We showered and washed our clothes, then borrowed all the irons in the hotel to try and dry them.

At 10.30 we sat down, a little damp, to a massive meal in the a la carte restaurant and celebrated, perhaps prematurely, our escape.

The next morning we awoke late. I was stiff and tired from the previous day's marathon drive and my hand was troubling me. Luke went to the local PAC headquarters while Jan and I ate a leisurely breakfast, worrying about Luke's future. When he came back, Luke announced that transport had been arranged directly to Elizabethville, so regretfully, he would be leaving us.

Although we had only known each other for two days, we said our farewells like two long lost friends. Luke took my gun and ammunition and walked off, newspaper bundle under his arm. His trousers now six inches too short, he looked a pathetic sight, but it had to be seen as a hero's departure.

We were on the road again, a little north of Lusaka Jan suggested he should drive. I was too tired to argue and just crawled into the passenger seat and took the unusual precaution of strapping myself in. I don't know why, I don't think I had ever done it before as in those days seatbelts were an oddity. I went to sleep.

I woke up disorientated. It was raining, but the rain was coming up from the ground and I was weightless.

It didn't take me very long to realise that we were actually upside down, we appeared to be in a ditch. After a struggle Jan got me out and we looked at the situation. It was not good.

Jan, for an intrepid freedom fighter was looking uncommonly guilty. The car was indeed upside down in a ditch. The windscreen and rear window had gone, the doors, roof and bonnet were dented and bent, generally in quite a mess.

'How fast?' I asked.

'About 80K,' he replied. 'It had been doing it for a long time, suddenly I couldn't hold it.'

Likely story, I thought, 'Doing what?'

Behind us the road was straight. I knew this part, it was not only straight, but straight for hundreds of kilometres. It was a wide gravel road, slightly raised in the centre and falling gently away to a ditch or storm drain on either side. In the rain, a little like driving along the top of an ice covered railway carriage. Tricky, but others had done it so why couldn't he?

I could see the skid marks weaving from one side of the centre of the road to the other side. I walked back down the road, the weaving continued for as far as I could see.

'How long has it been doing that?' I asked.

'About 50K.'

'50K! You've got to be joking?'

'No, really.'

'Why didn't you stop, didn't you realise there must be something wrong?'

'I thought it was the type of car and the mud. I have never driven a front wheel drive before.'

'Was he right?'

We turned the car over without much difficulty and found the engine and drive to be okay, but for some reason it wouldn't pull itself out of the ditch. Then I found the problem. Both rear wheels were locked solid with mud in the wheel arches. Jan had been driving along at 80 kph with both rear wheels dragging like a sledge. No wonder he lost control of the car.

We cleared the mud from the wheel arches and the car easily pulled itself out of the ditch and back on to the road. After repacking I took over the driving and we proceeded with caution. No problem except that without a windscreen we were battered with rain and insects, making 40 kph about the maximum speed possible.

We stopped at an inn for some food. The proprietor was not too happy to see Jan but served us begrudgingly. Only two weeks, I thought. No doubt they must have thought the same thing because the next time I passed through they had gone. Independence cuts both ways.

I found a narrow piece of perspex and strapped it on to form a rain shield. I couldn't see through it but by alternatively sheltering behind it and then peering round the side to see the road, we could make better time.

Suddenly a figure appeared from behind a tree and waved us down.

'That must be my man,' said Jan. 'We must be very close to the border.'

Sure enough the remarkable bush telegraph had worked. We were hundreds of kilometres north of Lusaka, there was no telephone in the vicinity and yet on the same day we had left the capital there was

21

a guide waiting to stop us before the border. Jan got out ready to spend another night walking through the bush. We arranged a rendezvous point for the following morning.

I also spent a cold and wet night in the windscreenless car, but passed through the border without any problems. The officious Scotsman had gone and the new immigration officer laughed at the Prohibited Immigrant stamp.

'You won't be needing that for long,' he joked.

'How do I get rid of it?'

'Just get a new passport when you get back to the UK, old boy. All that nonsense is just about over here, won't last much longer down south either. Of course you'll have a new lot to deal with,' he nodded at his African assistant.

On the 'almost Tanzania' side the staff were mainly black. No-one recognised me or the car or knew that I had spent five nights in their jail some six months ago.

Five kilometres after the border I picked up Jan and we drove slowly through to Mbeya where I paused for a few minutes to inspect the wreck of a short wheelbase Land Rover I had abandoned in 1959. It was still complete apart from the spot lights and tyres. Nothing else was of much value on it apart from the nuts and bolts. After all, what can you do with an engine split open at the crankcase, two hopelessly bent axles and a disintegrated gearbox?

We climbed the escarpment after Mbeya in howling winds, torrential rain, making it virtually impossible to see. It was more like swimming through a storm than driving a car. The journey was becoming a nightmare.

Somehow we staggered into Iringa where Jan left me. He hitched a lift to Dar while I crawled on my way towards Nairobi. He was home and free, no more borders to cross – until the next time. For me there were still more than 1,000 alarming kilometres to cover.

That night, at about 10 o'clock, the right hand drive shaft broke, either due to old age or as a result of the crash. In the dark, just before a right hand bend on a corrugated road somewhere between Iringa and Dodoma. I removed the hub, took out the drive shaft, wedged a tyre lever in the remaining part of the universal joint, effectively locking the differential. I tied the tyre lever in position, refilled the brakes with water as I had no brake fluid and I was back on the road.

Driving on one wheel in a front–wheel drive car that does not have true zero scrub steering is interesting, especially on a gravel road. I put my foot down to turn right and lifted it off to turn left. After a while I became quite good at it, setting up the balance through the corners so that I did not need to turn the steering wheel.

In the dawn light disaster struck again. The tyre lever, which was wedged in the differential, had worked itself loose and begun to move around. It had broken off the cylinder block drain plug causing the engine to boil dry.

I found a river, collected water and a piece of wood to serve as a plug. The tyre lever was repositioned and I limped on to Dodoma where I decide to abort the trip and abandon the car at George's garage. At this stage the car was mechanically more or less in one piece, at least the engine and gearbox were fine. I doubted whether they would be should I continue through to Nairobi. I also had a plane to catch. I did feel a little guilty but my endurance was wearing thin.

So I took one of the more expensive cabs of my life, flew the same night to England and wrote a long explanatory letter to Benbros Motors, telling them where their car was, the state it was in and apologised profusely.

It was during that last night, battered and exhausted, driving on one wheel, that I coined the name AFRICAR. I knew that I did not need, nor did Africa for the most part need, a Land Rover or Land Rover equivalent. I had tried it and found it wanting. I also knew the standard consumer car was not the answer.

What I needed, desperately at that moment, was a car built for, and preferably in, Africa, by people who understand the local conditions and priorities. A vehicle quite different from those dreamed up in Detroit, Tokyo, Wolfsburg or Turin. Perhaps something a little closer to those thought out in Paris.

What I needed was an AFRICAR!

3
ARCTIC

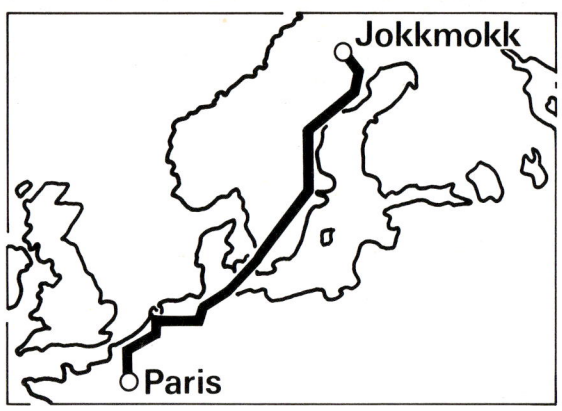

In the years between 1964 and 1978 I pursued a career in photography and film making. My interest in vehicles was strictly as a consumer, but the nature of my work provided accumulated driving experience of some half a million miles in Europe and America and about the same distance in the developing countries.

It was a period in which I learnt to appreciate the sophisticated engineering thinking behind a whole range of French vehicles from the humble 2CV and the Renault 4, to the big DS Citroens. In Africa, I also learnt to appreciate the enduring simplicity of Peugeots which really put to shame almost everything that had come before them.

It was not until an unusual incident occurred in California that I again became interested in the actual construction of vehicles. I was using a Citroen Méhari, a jeep-bodied, plastic panelled 2CV, at the time. On the 4th of July 1976 it was parked down-town Los Angeles outside the Wells Fargo Bank while I took some photographs from a roof of a nearby building. Plumes of smoke and fire engine sirens suggested to me that something might be happening in the street. And yes it was, the Méhari had caught fire. According to bystanders some kids had thrown fire crackers into the petrol tank. An interesting American way of celebrating independence.

The car had already set the bank building on fire up to about the fifth story. That evening Carolyn Hicks and I scraped the remains into the back of a tow truck and took them home to the Hollywood Hills. Just two weeks later we had planned to go to a Citroen Car Club meeting at Santa Barbara. I decided to rebuild the Méhari from plywood, glue and polyurethane, the only materials I had to hand.

Exactly 14 days after the car had burnt out we arrived in Santa Barbara after a trouble free trip, apart from noise and wind on the American freeways. That was the beginning of a realisation that a stressed plywood structure could be a very suitable way to build a car. A year later, after looking at a catamaran design, I became interested in the cold-moulded wood and epoxy technique. It was a gradual understanding of what epoxy could do for plywood thay finally triggered my decision to build what had already been established in my mind as an AFRICAR.

There was one more step. For many years my photography and film making had been in the area of comment and criticism, I had spend 20 years taking an analytical approach to the world and had frequently found that my experiences, particularly in the Third World, were at odds with the Western perception of both conditions and requirements in those countries. I felt a need to do something positive.

The process that led to AFRICAR started with a concept for a television series critical of the consumerisation of the motor car. But instead of just being critical, I decided to take a positive step and build a vehicle that embodied the ideas I knew were essential for worldwide use on the ordinary road - and worldwide manufacture within ordinary economies.

Parts of my first wooden car based on a Citroen 2CV chassis ▷

My first wooden car built in California in 1976 with plywood and polyurethane, used for two years on U.S. freeways ▽

◁ **The Arctic road near Jokkmokk on the Arctic Circle**

△ **Our holiday cabin base at Jokkmokk**

It's February 1984, we have just covered one thousand five hundred kilometres in 19 hours on ice and snow. We are approaching Jokkmokk, a small Swedish town situated right on the Arctic Circle. Temperatures at night are down to minus 30 degrees centigrade.

The theory is that a vehicle designed to really cope with all African conditions will work that bit better than most in other parts of the world.

So we are heading north of the Arctic Circle for a few days of cold weather testing before setting out on a 30,000 kilometre proving run to the Equator.

To me it is amazing and it feels good to be driving along the roads and tracks of the Arctic in a vehicle I first thought about while stuck in the mud of Northern Zambia 20 years ago.

Driving through the Arctic is also tainted with nervousness. It seems a long way to Nairobi and I know it could be considered foolhardy to be running prototypes on their first extended road test over such a difficult and lonely route.

Our cars have an advantage. Their design

Herds of reindeer roam free in the Arctic
▽

started not as an exercise in engineering or marketing, but as a research project to investigate the compromises that are inevitable in every attempt that we make to ease our burden of travel and transport.

Here in the Arctic huge herds of reindeer roam free. They are classic beasts of burden having been harnessed and put to work for thousands of years.

Snow–mobiles, complex machines with

27

The pick–up won't start because of badly adjusted plugs in Southern Sweden

A typical bushsledge used in the Arctic region of Scandinavia

infinitely variable transmission, are effective, but on long and arduous cross–country journeys they are still no match for dog sled teams.

To really understand the compromises of transport we must go back to the fundamental – walking.

People are not naturally well adapted for walking on snow and ice so in the northern regions there are snow shoes and skis, snow boots and 'perambulators' – simple push sleds which must surely deserve a place of honour in any transport museum.

Despite our limitations on snow and ice, we come from an illustrious lineage of walking ancestors.

Paleontologists tell us that the first human beings evolved or appeared somewhere in central Africa. It is possible that this theory could change with new evidence, but whether we derive from one or several sources, our forebears colonised the

△
A forest ranger using a snowmobile near Jokkmokk

Niger ▷▷

entire planet using little more than their feet to complete the earliest migrations.

While we walked no region was inaccessible to us and down the years we've overcome almost every obstacle. People, with a little help from their hands and their friends, have walked to the top of Mount Everest, and people, with a little help from Nasa, have walked on the Moon.

Today, in our mechanised world, far more miles are still covered on foot than by any other means. It is the next step that really starts the compromises.

The first utility vehicle or pick–up was, it seems, the first person who was persuaded to carry a massive load on head or back.

The first motor car, that is a self–propelled personnel carrier, was undoubtably a woman, carrying a demanding if unskilled driver, in the form of a child.

In this first stage of assisted transport mobility is lost and that must be set against the advantages gained – by some.

Animals, from horses, donkeys and mules, to water buffalo, oxen, camels and elephants, have brought us many advantages. But we have had to pay.

Animals need breeding, taming, training – they need harnesses, sleds and carts. Animals have a will of their own and have to be fed and watered. They need skilled drivers and handlers and are, surprisingly, often less agile than us.

With animals we gain load carrying and we may gain speed but we actually lose mobility, flexibility and the costs are high.

The wheel, which evolved after various forms of the sledge, is probably not more than 10,000 years old. It may have developed from rollers, such as tree trunks, placed under heavy rocks while moving them into position for buildings and monuments. More likely, the wheel comes from

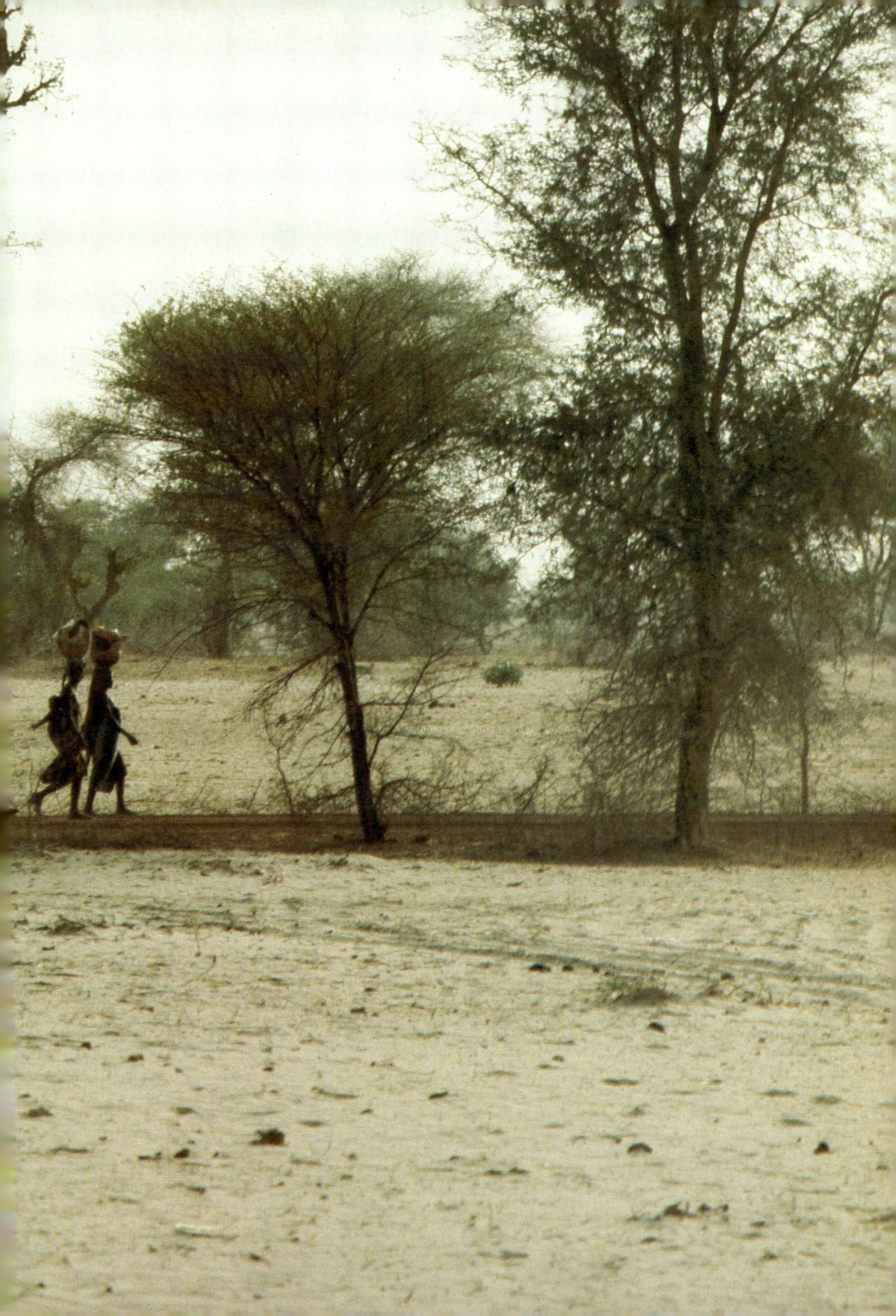

Horse–drawn taxis in Zermatt

the pulley developed to help draw water from deep wells.

The wheel on its own is pretty useless, it is the axle and the bearing that make it useful and all these elements are apparent at the well head.

Another plausible theory is that the wheel first evolved from the potter's wheel. Certainly pots were being thrown on wheels long before there is much evidence of wheeled traffic.

There are times when the modern traveller can be forgiven for wishing that the wheel had remained at the well head or with the inventive fingers of the potter. For the wheel, seen as such a liberating concept for machines, represents the biggest compromise to date in all–terrain transport.

The wheel is truly unnatural. It suits very few of the ordinary surfaces which we may have to travel over. The wheel requires extraordinary roads or tracks to be built for it if it is to work efficiently or, in many places, if it is to work at all.

The wheel has possibly done more to divide the world into supposed accessible and cut–off regions than any other invention, and it still divides us.

In fact the wheel is best when it hardly ever touches the ground.

Powered flight, not the motor vehicle, has become the expensive 20th century antidote – opening up the world again and bringing communities, isolated by the wheel, back into contact.

Here amongst ice and snow but on specially constructed roads, the wheel seems to be in its element. But the edge is very close and quite unforgiving, even with six wheels.

Our overland route will amount to a distance equivalent to threequarters of the circumference of the earth. Further south, in some places, roads will not exist at all. If the design is right, our cars for Africa will overcome the many day to day problems of travelling with wheeled vehicles.

The team has come together almost by accident. To test the cars under consumer conditions, no one has any special driving

△
Deep well in the Southern Sahara near Agadez

Four-wheel drive in Niger ▷

skills.

Liza Mellor has driven from London to Turkey and has travelled in the Soviet Union. She is the cook and the main driver of the pick-up.

Tony Hughes, Africanist and journalist, is to deal with formalities and smooth our way across frontiers.

Bob Williams asked to work on building the prototypes in order to come to Africa for, as he said, a last grand adventure. He is the mechanic and driver of the six-wheeler.

Apart from myself, Carolyn Hicks is the only member of the team who has been involved in AFRICAR development from the beginning. She is provisioner, trouble-shooter and driver of the wagon.

CAROLYN: 'When we were driving on ice near Jokkmokk, I didn't know it was ice – the car just went straight, it seemed to be handling well and moving along nicely. It was only when we got out of the car and all promptly fell on our back ends because we couldn't stand up on the road that I realised.'

△
Four-wheel drive in Niger

Charles Best also helped build these prototypes. He is with us in his professional role as a photographer.

CHARLES: 'The pick-up's great. It's the lightest, the fastest, the most positive and the best fun to drive. You get in it in the morning and feel like going for a spin or just having a play.'

33

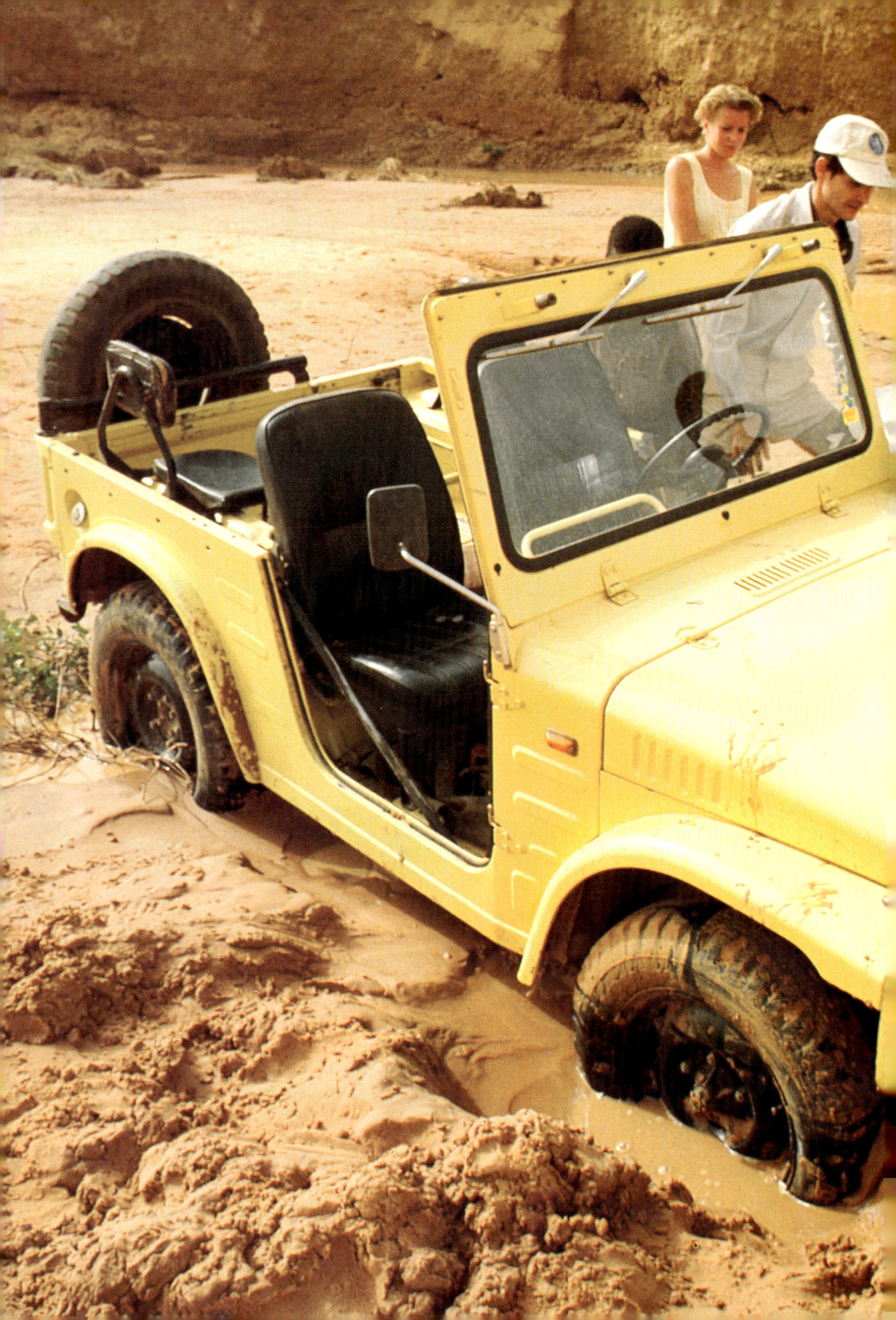

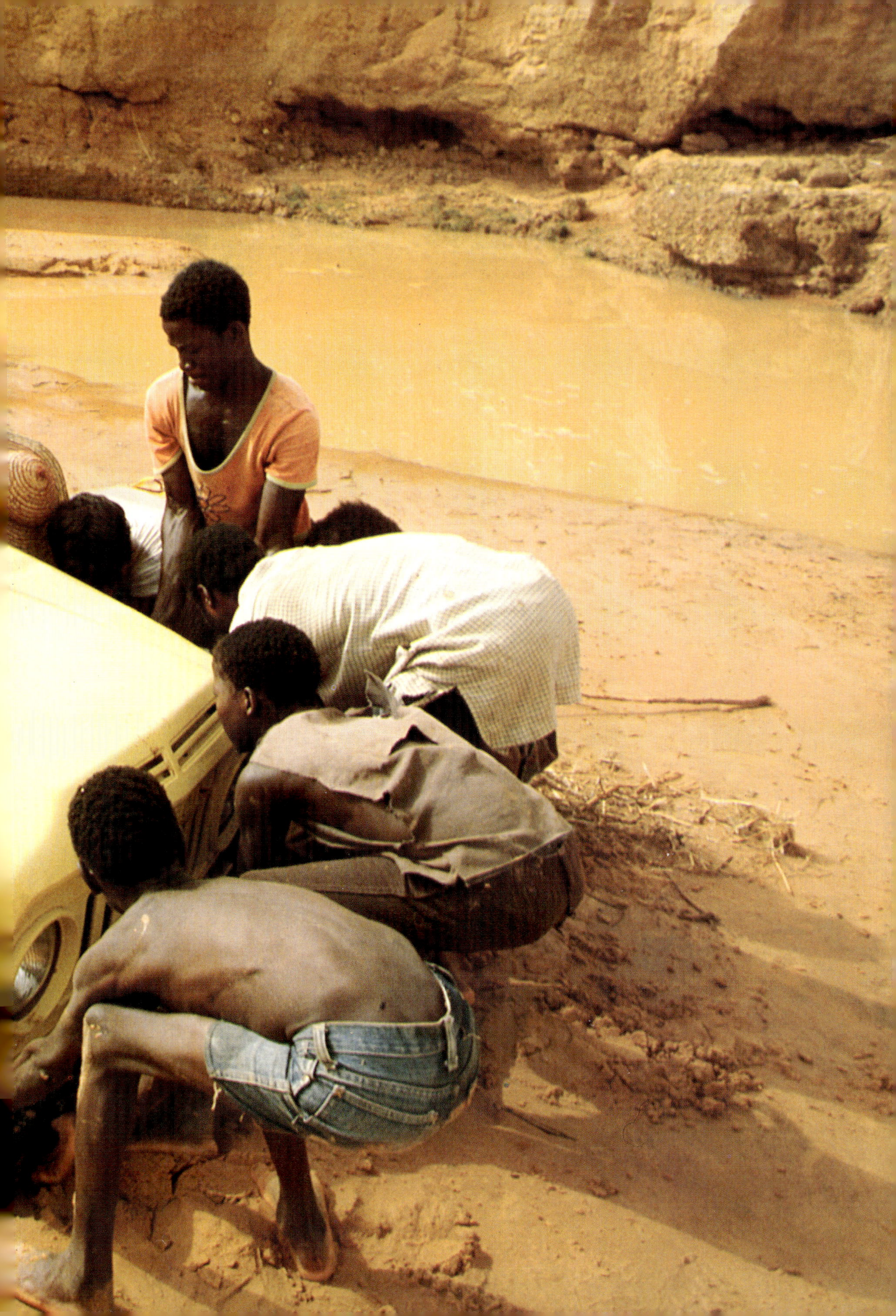

I have to hold the bits together – both mechanical and human and see we all arrive safely in Kenya.

I wanted a vehicle that could be driven five hundred or a thousand kilometres without battering the contents. Last night was the best example so far when we did all our work during the day, got into the vehicles at half past five, drove 550 kilometres then went to bed. On the whole, cross country vehicles that I have experienced just don't like doing that.

△
The result of parking off the specially built road in the Arctic

LIZA: 'I like the lightness of the pick–up. I think the six–wheeler is more like driving a lorry and as Bob prefers this, he really enjoys the six–wheeler. We had a cup of tea perched just behind the brake for about five miles and didn't spill a drop.'

BOB: 'The six–wheeler rides like a coach over the bumps and humps, it's really nice.'

TONY: 'I feel the need for more places to put down a book, paper, a briefcase or even a pen, to put the cup – oh, for the cup of tea we have a good place.'

It would be nice to have a proper gearbox and of course the great tragedy, in a way, is that right at the last minute we found the connecting links in the four–wheel drive system were not up to it. So, we are now running front–wheel drive only on all vehicles.

Starting from Jokkmokk we intend to drive down through Scandinavia and Europe to Sicily.

After the ferry to Tunis comes the real test of a long and tortuous route across the Algerian Sahara before heading south again through Niger to Nigeria.

The AFRICAR team at the Arctic Circle. L/R: Liza Mellor, Charles Best, Anthony Howarth, Tony Hughes, Bob Williams, Carolyn Hicks

At Kano we turn east to the Cameroon Republic, the Central African Republic, Zaire and Uganda before reaching our final destination, Nairobi, capital of Kenya.

Such are the best laid plans, in practice it worked out a little differently. Africa has a way with plans.

In 1959 I drove round Africa in a Series 2 Land Rover, a vehicle very similar to the more modern Series 3. It was a journey of 40,000 miles covering about 30 of the 42 countries on the African continent, taking about a year.

That first journey was a success. My travelling companion Peter Turner and I made it back to England in one piece, with most of our vehicle.

The vehicle itself? Well, it was a long time ago, shortly after World War II that Land Rover cleverly adapted the American military Jeep concept to practical civilian use. In a sense almost all modern four–wheel drives are 1949 Land Rovers, or perhaps 1940 Jeeps, under their gaudy colours and go–faster stripes.

But in 1959, in Africa, our Land Rover showed little sign of the ten years development for the conditions that we met on normal African roads.

We broke axles and springs, shock absorbers, gearboxes, clutches and door locks. Mostly items that one might expect to break on rough roads in a vehicle with solid beam axles and sprung like a 19th century cart. But then that was the problem – Newton's laws of motion hold even for 4x4s.

Three years after the Land Rover trip I made a second long African overland journey, 27,000 miles or 40,000 kilometres in six months.

For this trip I chose a small family car, the then newly introduced Austin 1100 – it was good to be back on the open road in Africa.

The Austin, a thoroughly modern design in the French tradition with fully independent suspension and front–wheel drive, was impressive.

The ride and visibility were good and the adhesion exceptional. Averaging relatively high speeds and overloaded, the Austin usually got me to my destination.

Me with Peter Turner on our Landrover in Cape Town, 1959
▽

△
The Austin 1100 with which I travelled in Central and Southern Africa in 1963

Me with the first prototype AFRICAR, August 1982
▽

As a car for Africa the 1100 lacked ground clearance and particularly clearance in the wheel arches. It was too fragile, although easily repairable when it broke. It certainly taught me the value of independent suspension and dominant front–wheel drive for African conditions.

It was on this second African journey that I coined the name AFRICAR. I had discovered that I and Africa for the most part did not need a Land Rover nor did we need a Western style car suitable for travelling on tarmac. What we needed was a vehicle that used all available technology. The kind of technology which is so often ignored by the major car manufacturers, to make it effective both on and off the road. A lightweight, multi–purpose vehicle – an AFRICAR.

That was over 20 years ago, since then very little has changed.

It was the sheer frustration with that lack of change in all-terrain vehicle technology that led me, in 1979, to take a clean sheet of paper approach to the design of an AFRICAR.

But the real excitement was the vision of genuine local vehicle manufacture easing many of the problems of a hundred and forty or more of the world's poorer countries.

Meanwhile, in Southern Sweden we have our first problem, the pick–up won't start. It turns out to be nothing more than cold weather and maladjusted sparking plugs.

Each AFRICAR has twin fuel tanks holding the total of 110 litres required to give us the minimum range we will need in Africa, well over 1,000 kilometres.

Watching over 300 litres of refined unrenewable oil products disappear into our cars, I couldn't help but feel that the old concept of putting the world on wheels has a limited future.

In due course there has to be a better way. After all, it was only recently that the advent of mass private car ownership made barriers out of the world's natural waterways.

Up to less than a century ago, rivers and oceans were roads and freeways. Far from being barriers, rivers were the easiest

Sydney Harbour Bridge, Australia
▽

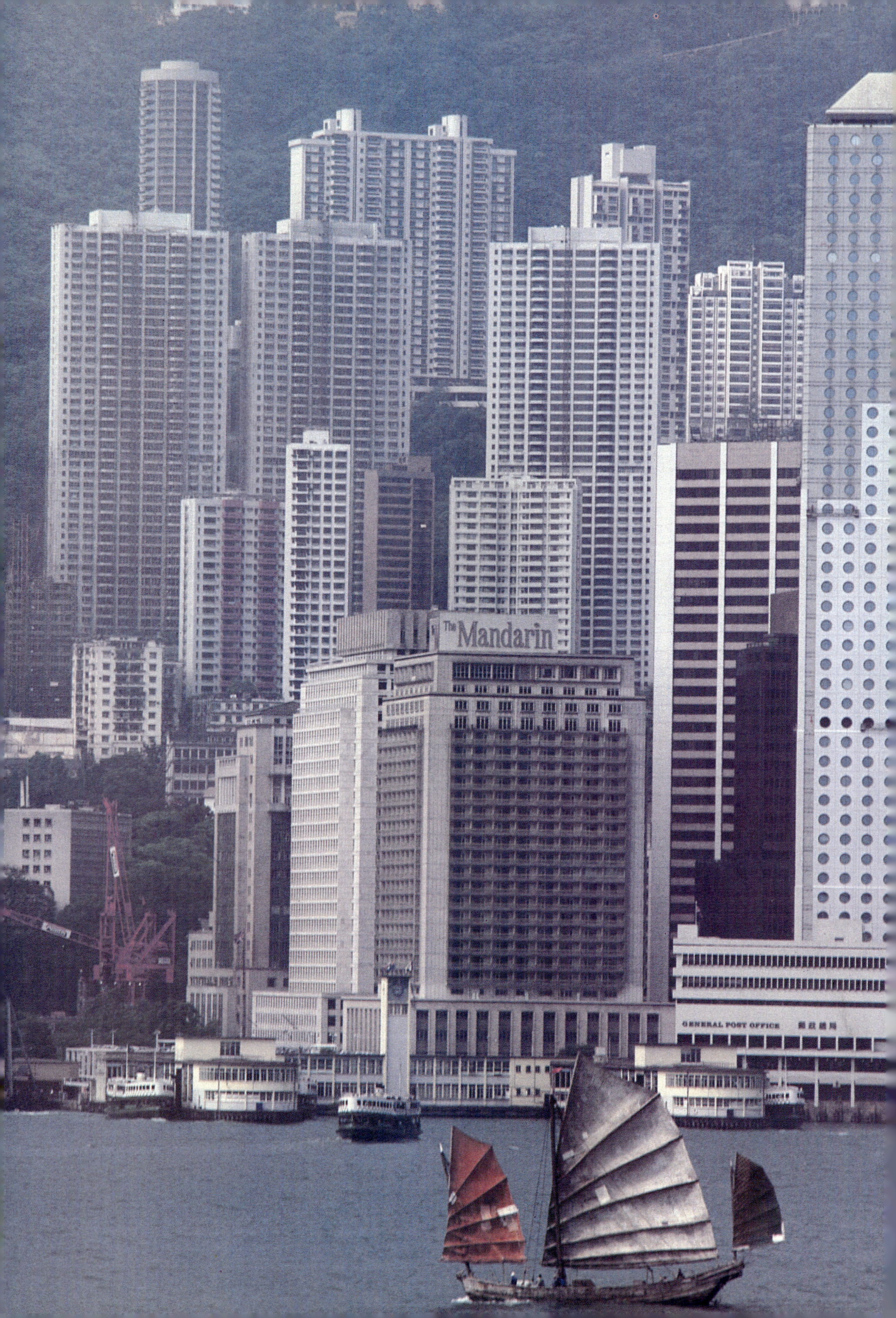
The Mandarin
GENERAL POST OFFICE

◁ **Sailing junk in Hong Kong harbour**

△
Building traditional sailing junks using modern power tools

The coastal plain of Sarawak in Borneo ▷

means of travelling inland and oceans provided access between continents and peoples.

The location of cities and the distribution of populations throughout the world were almost entirely dependent on waterways, not motorways.

Today most inter-continental passengers travel by air, while bridges have been built where possible and where they can be afforded. But the oceans are still the freight routes of the world and water-borne communities flourish.

Boats with sails, or wind motors, were the first form of self-propelled transport used by our ancestors. Every region of the world evolved a sophisticated sail system to suit local conditions.

Sail power is still widely used in the Middle East and Asia. It is only in the Western waters that the sail has been relegated exclusively to the leisure industry.

Meanwhile the art of building robust commercial sailing junks from local timbers has not been lost. Simple modern power tools ease and speed the operation.

An efficient combination of appropriate technologies giving clues to what could be possible with more emphasis on local manufacture and less reliance on the products of centralised industry.

In Sarawak on the island of Borneo, rivers are the only practical highways. The jungle of Borneo is dense and the land rarely flat. There is a series of hills climbing into the interior with slopes of about 60 degrees, creating impossible terrain for vehicles or road building.

Here over 200 direct jungle miles from the capital city, the Iban live in long houses built on stilts beside the rivers. These houses may contain a hundred families and resemble modern tower blocks laid on their sides. There are private apartments as well as communal working and recreation areas.

Life is, at first glance, idyllic, primitive and remote. Yet the Iban have never been cut off from the world. They are also known

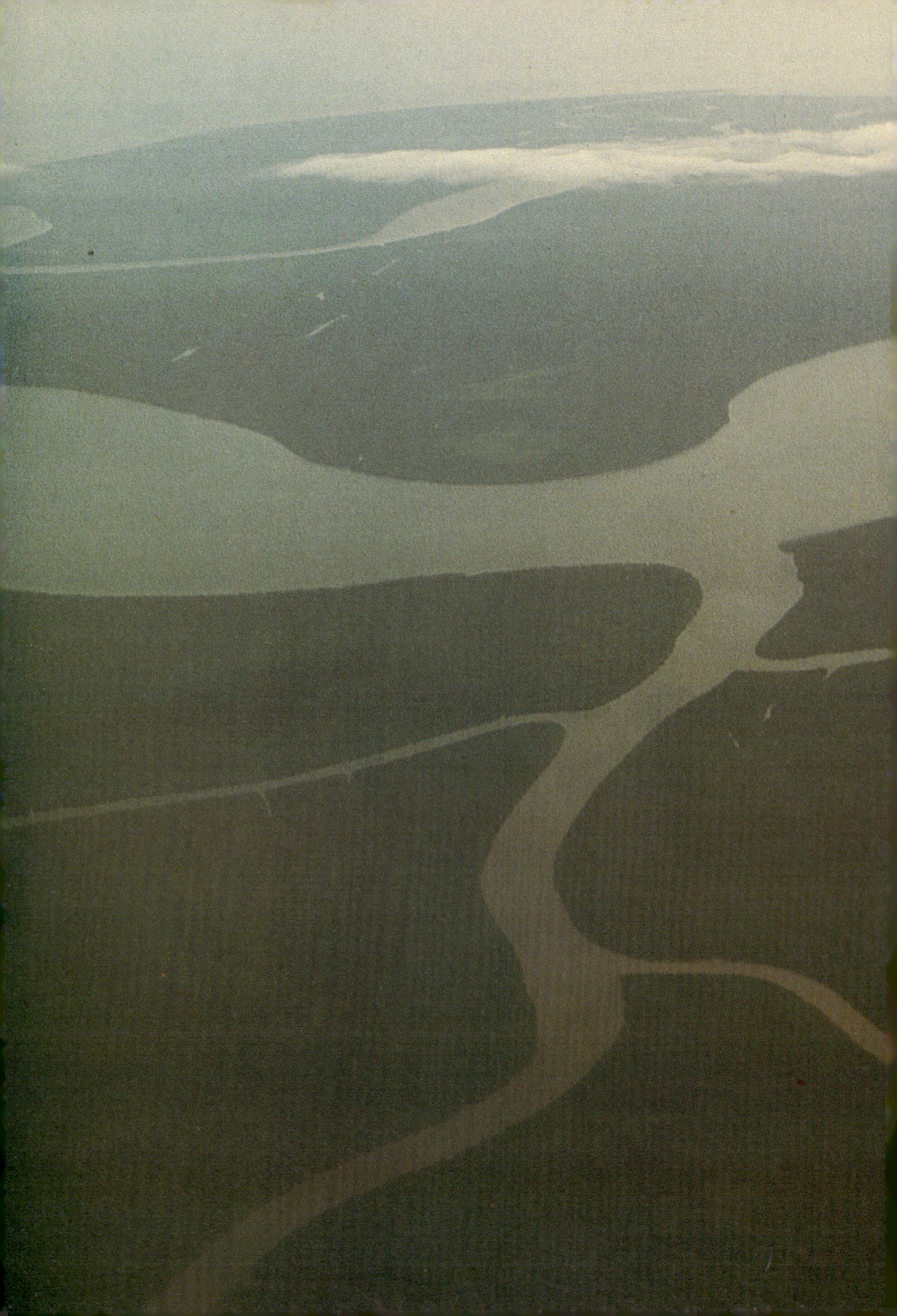

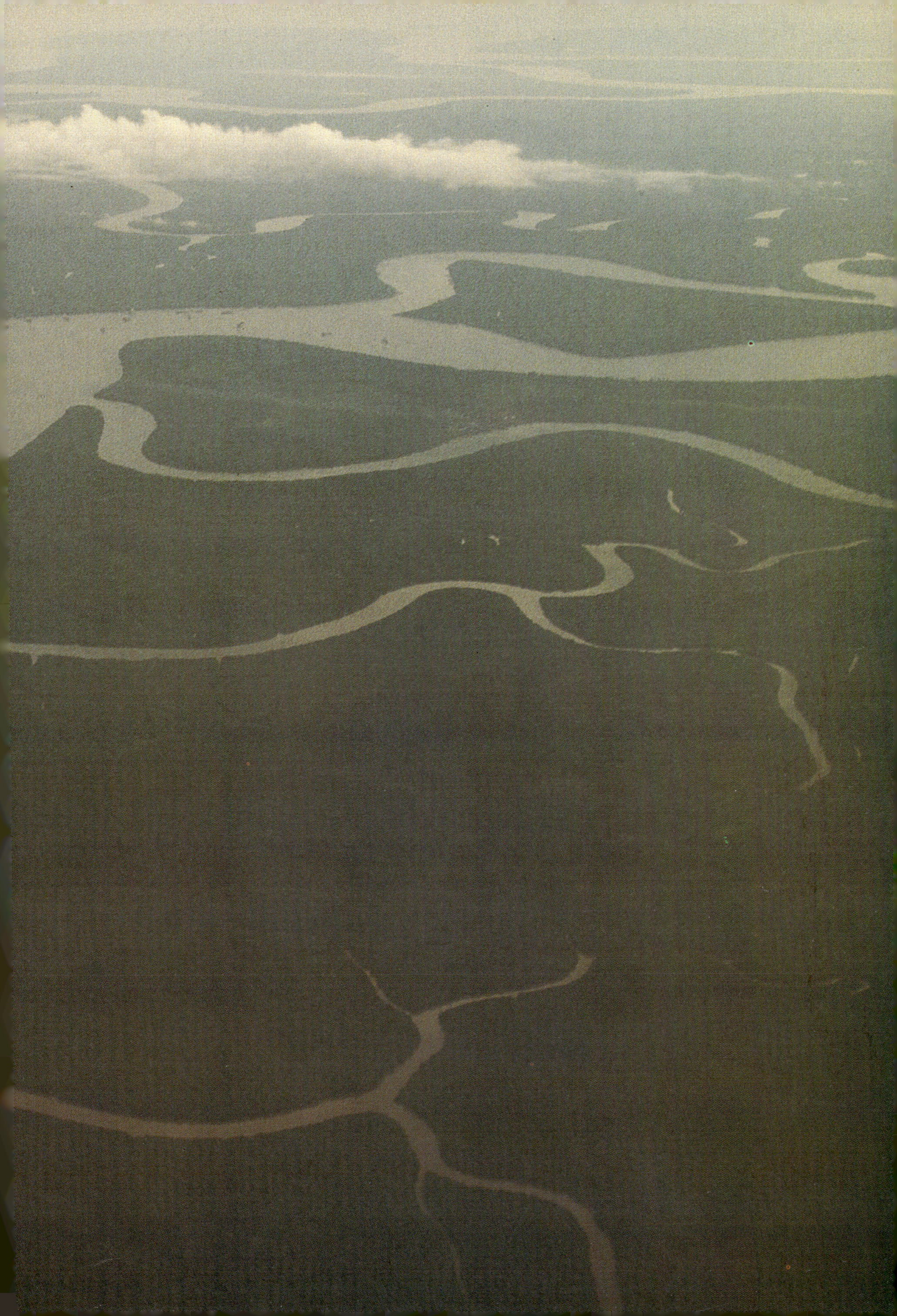

Iban long house on the Bangkit River

Iban travellers on the Bangkit

The Iban build their own boats called Prous

Motorised Prou ▷

as Sea Dyaks, once famous as coastal pirates and infamous as head hunters.

The Iban build their own distinctive boats. Called Prous, they are either carved from a single tree or built up at the sides with ribs and planks. The structure is sealed and reinforced with fillets of natural resin in the same way that modern hi–tech wooden racing boats are reinforced with fillets of synthetic epoxy resin.

Anyone who doubts that the use of boats and the mastery of water transport is as old as the human race, need only spend a few minutes beside a river in the interior of Borneo. Today it is the big outboard engines that gives the clue to contact with the outside world. From the houses along the Bangkit River, there are young men who work in oil fields in Brunei, in the Gulf and as far away as Scotland and Texas.

They appear to think nothing of coming home for a long weekend, for a festival, a wedding or just to party. Then after a few days they are off again, on 'bijoulai' or

Running the rapids on the Bangkit

walk–about, to places and cultures which, although alien, are familiar after years of contact.

Like all seasoned travellers, the Iban

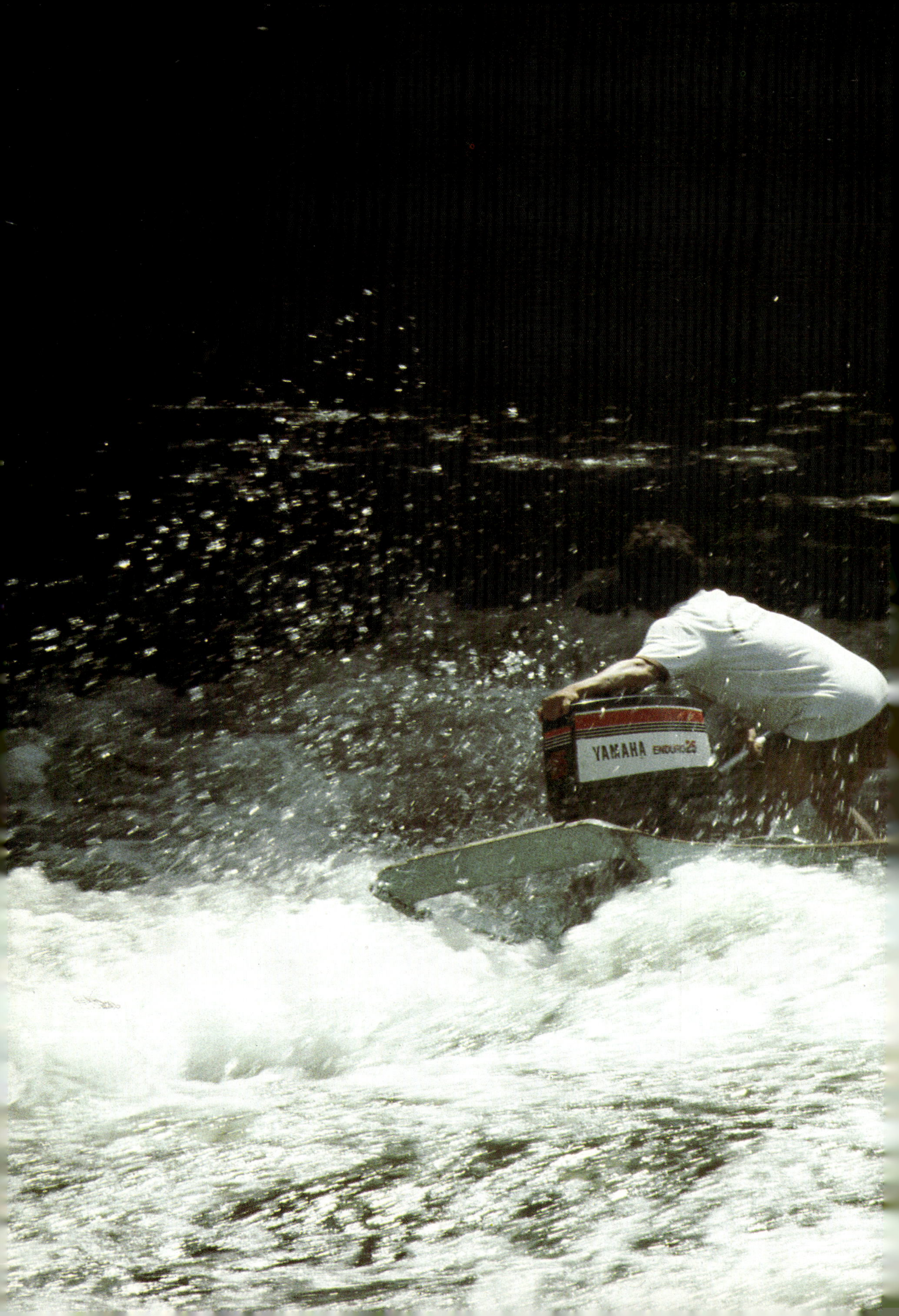
YAMAHA ENDURO 25

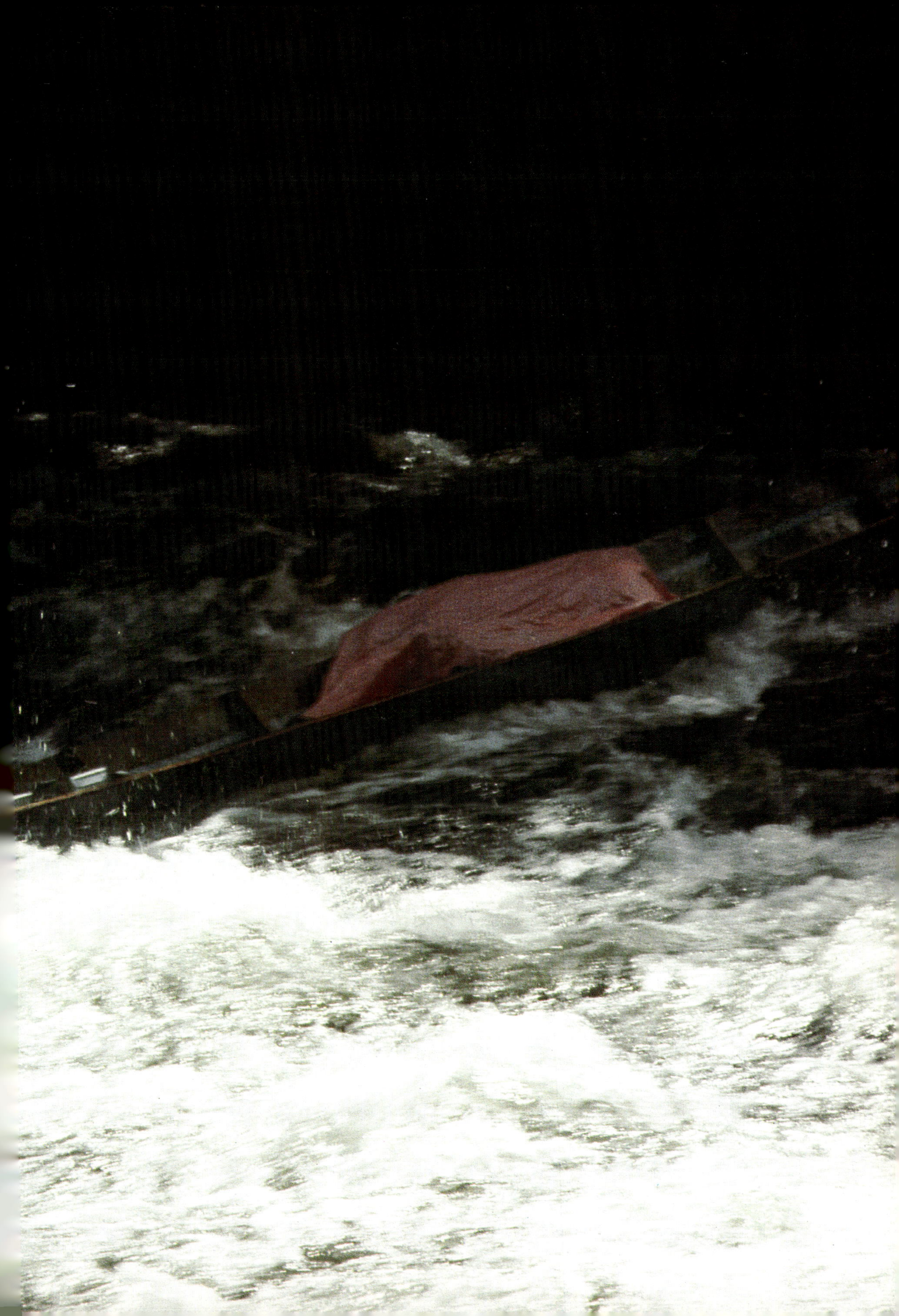

seem to leave for the airport at the last minute. However, there are more than 300 kilometres of rocks, rapids and water to cover before it will be time to fasten seat belts and order a drink.

In this country five miles a day is considered good going for an army patrol on land. A bus can take 10 hours to cover a straight line distance of 100 kilometres on one of the few roads. The river network is always faster. As the water and especially the rapids get deeper it is amazing what 25 horsepower of Japanese two–stroke motor can do to a 12 metre long pencil–thin, lightweight boat.

The Prou is the right vehicle in the right environment.

The Bangkit joins another larger tributary. Now it's full throttle all the way to the main river. Speeds point to point of 30 miles per hour are quite practical down stream. Up stream ten miles per hour or more is possible.

The one drawback of the motorised Prou is its dependence on typically inefficient mass–market outboard engines.

At Song, a small Chinese trading post, the Iban travellers board one of the Rejang River expresses.

Now we have multi–cylinder caterpillar or Rolls–Royce engines with hundreds rather than tens of horsepower. Speed

◁ Riverbus on the Rejang River

Riverbus passengers on the Rejang River ▽

seems to be the idea but schedules include picking up passengers from landing stages attached to their houses as well as from the villages and towns along the way.

Borneo has the great advantage of having a ready made natural highway system. In Africa there are few navigable waterways in relation to the size of the continent. The wheel is the only option – but the clumsy lumbering gait of the Saharan public transport makes a poor comparison with the speed and elegance of river buses.

At Seriki, close to the mouth of the river, we are in time to catch the air–conditioned Concorde. Everything goes on board, notably wheeled transport being moved from one small infrastructure of city roads to another.

Dodging escaped logs floating in the coastal waters, the last stage of the trip to Kuching is completed in a few hours.

Many of the Iban travellers are not stopping here, their journey continues by air.

It is possible to travel from the long house, Rhuma Langa, on the Bangkit River, to Aberdeen, Scotland, in about twenty four hours – if you time it right and the propellor shear pin doesn't break on your outboard.

Back in Scandinavia, our ferry journey from Halsingborg in Sweden to Helsingor in Denmark took less than an hour.

In Denmark we make the transition from

△
The Hammel Car of 1886

Scandinavia to Europe. It was in Europe that the motor vehicle, still for many places the only serious option for overland transport, originated.

After the sleek jungle Prous, the first horseless carriages are a bit of a let down. Spindly, wobbly machines chugging away at little more than a walking pace. Smelly, noisy, maybe unreliable and hardly safe – not a lot has changed.

The Hammel car was built in Denmark by Hans Urban Johansen for his employer, Albert Hammel, in the year 1886. The Hammel is arguably the oldest internal combustion engined motor car that is still both original and practical to drive on the road. The Hammel is positive, has authority and really works, yet who has heard of it?

There are many pretenders for the title of 'inventor of the motor car'. The honours usually go to Benz and Daimler in Germany in 1885 or 1886, said to be the same year as Hammel. Benz is reputed to be the first to make motor cars to a fixed pattern and to offer them for sale.

Thus, with the motor car, we don't acknowledge the origin of technology. Instead we are expected to celebrate the first car salesman and the birth of a consumer industry.

The first motor vehicle actually predates Benz, Daimler and Hammel by over 115 years. It still exists and we will find it when we get to Paris.

In Amsterdam the AFRICARS are expected for their first major press conference.

INTERVIEWER: 'What's your basic philosophy?'

Primarily it is to build a car which could be manufactured in Africa, not assembled from a kit. It has to be manufactured in small numbers in an African country because they do not have the foreign exchange, the hard currency, to buy imported cars. Secondly we aim to make a car which will work under those extreme conditions, so it is really two things.

INTERVIEWER: 'What makes it so easy for

△
Press conference for the AFRICARS in Amsterdam

this car to be built in Africa itself?'

It's largely because of the construction with plywood and resin that we use, very similar to a boat building technique.

The great car makers together constitute by far the largest manufacturing industry in the world. In the mid–1980s they make around 40 million motor vehicles a year.

The total number of vehicles in the world is about 400 million and the annual related turnover, including running costs, reaches at least five thousand billion dollars. The actual figure may be far higher but the elaborate tentacles of the vehicle industry are almost impossible to untangle.

Yet after a century of the motor car and more than two centuries of motorised transport, this massive industry is extraordinarily parochial.

Today the population of the Western industrialised countries, about one billion, a fifth of the world's inhabitants, owns 90 percent of all motor vehicles.

If the Eastern bloc is included, the figure rises to 98 percent.

This leaves just two percent of all the world's motor vehicles, to serve the needs of three and a half billion people living in more than 140 countries.

Amsterdam

Ironically and perhaps tragically, these same countries are the primary source of the cheap raw materials and energy that feed the motor industry, and the use of motor vehicles.

The irony goes even deeper. Since 1970 the motor industry has shown little significant growth. The established market is already saturated, perhaps super–saturated.

The world's biggest industry is largely bankrupt of new markets, money and ideas. It's management appears to make no effort to develop product or production techniques suited to the one great growth market still available – the developing countries.

The concept behind AFRICAR is based on operating largely in local currencies, with small plants producing appropriate vehicles in numbers suitable to local markets.

This approach should not threaten the traditional car industry. They have defined their cake by excluding the countries and peoples that a car for Africa could serve – about threequarters of the world.

The Dutch have one of the best records of supplying development assistance. Considering the other critical needs for the Third World, I was surprised that they were so supportive of the concept of cars for Africa.

In their own country they manage very well with an intricate network of trams, subways, buses and water buses. Amsterdam could be one of the world's great pedestrian cities, but Holland is flat and this supports the device that could be called to the defence of the wheel as an aid to transport. The ordinary humble bicycle.

In Holland, where more people can afford to own a car than in North America, the bicycle is one of the many available alternatives. In Africa the choice is limited. Affordable and suitable transport is desperately needed.

On the opposite side of the world from Holland is the country that has become the world's largest motor vehicle manufacturer and exporter. Japan now makes over three million more vehicles a year than the combined total of the American trio, General Motors, Ford and Chrysler.

From their home factories, the Japanese auto makers export about six and a half million vehicles a year – ten times as many as the US based American companies.

Surely in Japan people commute to work in glossy cars, surely the bicycle and the variety of public transport must be long gone?

It seems that the bicycle, the pedestrian and public transport are not only alive and well in Japan, they are the nation's transportation systems.

In the post World War II period many Western countries have either deliberately, or under the influence of market forces, run down public transport in favour of the motor car.

In Japan the government policy has been the opposite. Public transport has been ex-

△
The defence of the wheel

Japan
▽

▵
Monorail in Tokyo
▿

panded and adapted to the needs of an affluent, busy people.

Not content with one of the world's most extensive underground train networks, coupled with superb suburban and local rail systems and the expanding use of monorails, the Japanese have taken the most outrageous step of all in high speed, long distance overland transport.

The bullet trains, in service since the early 1960s, are not just comfortable, quiet, high speed trains, as their imitators are elsewhere. They are the direct extension of the urban and suburban train systems.

On all the main line routes serving the whole of Japan, a bullet train leaves from Tokyo every twenty minutes.

You can book your seat or sleeping berth in advance or you can just drop into the station and catch the next train to your destination, which might be a thousand kilometres away, with the ease and often the frequency of taking the tube, the metro or the subway.

So where, in Japan, is the product of the world's largest industry?

Road traffic reflects a different balance to Europe and America, with more commercial vehicles than motor cars. At times, more trucks carrying cars than cars being driven. On the streets of central Tokyo there is

△
The bullet train. In service since the early 1960s

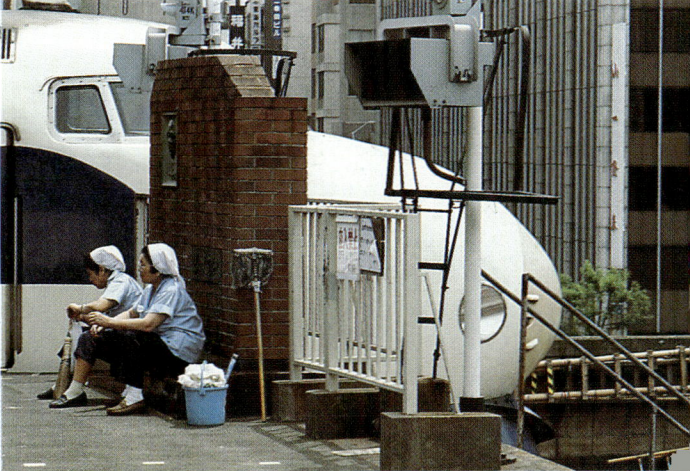

hardly a private car in sight, day or night.

In Japan, cars that have escaped export are sold, most charmingly, by the square metre of flawless paint work, by the quality and complication of the stereo system, by talking dashboards and the number of valves, turbos, intercoolers and horsepower.

Unlike its American counterpart, the Japanese car is not even needed for commuting to work. Little regard seems to be given to actual road performance, but absolute reliability is simply demanded and assumed by the customer.

At weekends cars and leisure vans are

△
Weekend traffic outside Tokyo

Lake resort near Tokyo
▽

taken for a promenade. They are to be found at picnic sites and motorway rest stops. Polished and gleaming with drivers and passengers dressed to impress, they are going nowhere special.

When the Japanese export drive gained momentum in the 1960s and 1970s, the home market had utilitarian tastes and needs. The vehicles that filled the ships were commendably reliable and well finished, factors that overshadowed their crude design engineering and often poor road worthiness. They were cheap and simple with spare parts remaining unchanged for years. They sold in any market that could be found. With no natural resources, other that an aptitude for organisation and work, Japan had to export or decline.

In the 1980s, with a Japanese vehicle exported every five and a quarter seconds, twenty four hours a day, seven days a week,

△
Cars awaiting export at the Nissan Dock in Yokohama

Nissan export ship
▽

Loading cars for export, Japan

365 days a year, the balance has changed.

Japan is now the world leader and so sets the fashion. The trivialisation of the light motor vehicle by the Japanese home market effects the world. It shapes the product of America, France and Germany.

The Japanese have now clearly decided to sell vehicles in Japan, roughly half their production, and then in the USA and Europe – in that order.

Like those that came before them, the Japanese now offer image and style, updated yearly, for use on tarmac. The rest of the world can either pave their roads or make do the best they can with the wild–west inspired off–road rider.

So when it came to designing the AFRICAR I knew that Japan was as unlikely a source of suitable vehicle technology as America would have been in the 1950s and 60s.

It was in France that I found the appropriate engineering tradition.

On the surface, rush hour in Paris seems chaotic, but don't be misled. For this is where it all began. Assisted by an intensely logical educational system that combines philosophy and the humanities with science and engineering, the French, not the Germans or the Italians or any of the others, wrote the book for the modern motor car.

I found that their solutions, often first used in Europe more than fifty years ago or more, could be equally applied to a car for Africa.

4
EUROPE

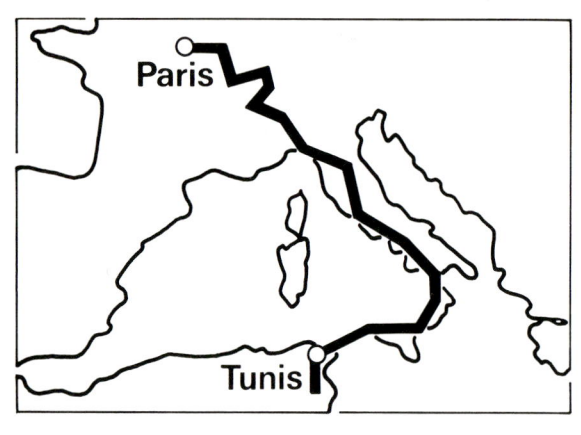

The motor car illustrates the gap between the rich and poor countries with more clarity than almost any other product.

When the roads of America were unpaved and for the most part poorly sealed, carriages had large diameter wheels to cope with the depth of mud.

Early motor cars followed this tradition, not as is often thought because they were carriages with engines in them, but because they needed the characteristics of the carriage wheels and suspension to cope with the conditions.

The Model T Ford of 1907 is a classic case, using the transverse spring, popular particularly on lighter carriages, to ensure the wheels stayed on the ground however uneven the surface. The moment roads became sealed with tarmac or concrete, motor vehicles 'developed' accordingly. Ground clearance was reduced and suspension movement decreased. In fact, the dynamic capabilities of the motor vehicle became closer to those of the go–cart, rather than those of a useful form of transportation.

The majority of the products made in the 1920s and 1930s, right through to those of the 50s and 60s, made little use of the simple but clever technology developed in the previous century. They simply had a couple of axles under a chassis which was 'knocked out' as fast and as cheaply as possible.

Even in those years the publicists of the motor industry managed to engender the image of high technology in the minds of

The world's first full size motorised vehicle built by Nicholas Cugnot in Paris, 1770
▽

the buying public. In practice, they were frequently selling something that the Romans would have recognised and certainly had little difficulty in understanding.

The challenge in developing vehicle technology is to make a vehicle that will function in all conditions and the basic systems for this have been around for decades.

There appears to be no logical reason why much of this technology was adapted to or developed for the motor vehicle in France. But that certainly was the case when it came to actual mass production vehicles. In other countries innovation usually ended up on limited–run specialist vehicles or never progressed beyond the prototype workshop. In France, from the 1930s onwards, genuine progressive vehicle technology was on sale to the public at a price frequently competitive with the cheapest the Americans could offer. It is possible that the vast small–farming community of France had a need for vehicles which really could travel across fields, particularly ploughed fields, carrying a basket of eggs without breaking them.

It is also possible that the close proximity of the French North African Empire, just across the Mediterranean, provided the French motor vehicle industry with immediate feed–back.

It seems more likely that the reason lies within the French education system which demands a philosophical approach, even to subjects like engineering and structures.

The fact is, that without the gambling innovation of Andre Citroen, the world would be a poorer place in vehicle technology. The motor car would probably not be the relatively safe, stable and predictable vehicle that today most of us drive on tarmac.

And proven examples of the geometric and dynamic principles required for the basic characteristics of the AFRICAR would, almost certainly, not have been available to us.

1922 Citroen with halftracks that made the first Sahara crossing
▽

1923 C–Type Citroen. Identical to the first vehicle to drive around Australia in 1926. This one also drove around Australia 50 years later in 1976, averaging about 30 miles an hour, doing 40 miles to the gallon and it broke just one light bulb

We have arrived in Paris, the home of the first motor vehicle. The motor industry, as we recognise it, was inaugurated and the book was written for both the traditional and the modern motor car in France.

'The first automobile in the world', as its plaque reads, is housed in a Paris museum. It is a strange looking vehicle. Built in 1771 by Nicolas Cugnot in Paris, it is in fact the second automobile in the world, as Cugnot built his first vehicle in 1769.

It was designed as a military tractor with a boiler at the front, two cylinders driving through a ratchet system and a massively heavy load on the single front driving wheel.

It steers through what can only be called a rack and pinion steering system, has quite

The Citroen Traction Avant 1934 to 1955, the classic modern motor car

The front end layout of the Traction Avant showing the front–wheel drive, torsion independent suspension and the interesting monocoque body form

Model T Ford – 1907 to 1927, over 15 million sold   [Photograph: Autocar magazine]

The 1935 concept which is still more advanced than most modern cars

Before the rest of the world had caught up, the Traction Avant was modernised into the Citroen DS

a good driving position and is fairly comfortable with excellent visibility. It has a rather amazing brake consisting of a sprag which digs into the tread on the front wheel. Behind is a huge load carrying space for guns, ammunitions and other supplies.

In the same Paris museum is a garish example of the world's first classic car, the Panhard et Levassor of 1894. The engine and radiator are in the front, the passenger compartment in the middle, a four speed sliding cog gearbox behind the engine with the drive to the rear axle. There you have it, basically the Ford Sierra of today!

By 1900, France had by far the world's largest car industry, and a component supply network that provided engines and

THE COMPOSITE SOLAR ELECTRIC DIESEL CAR

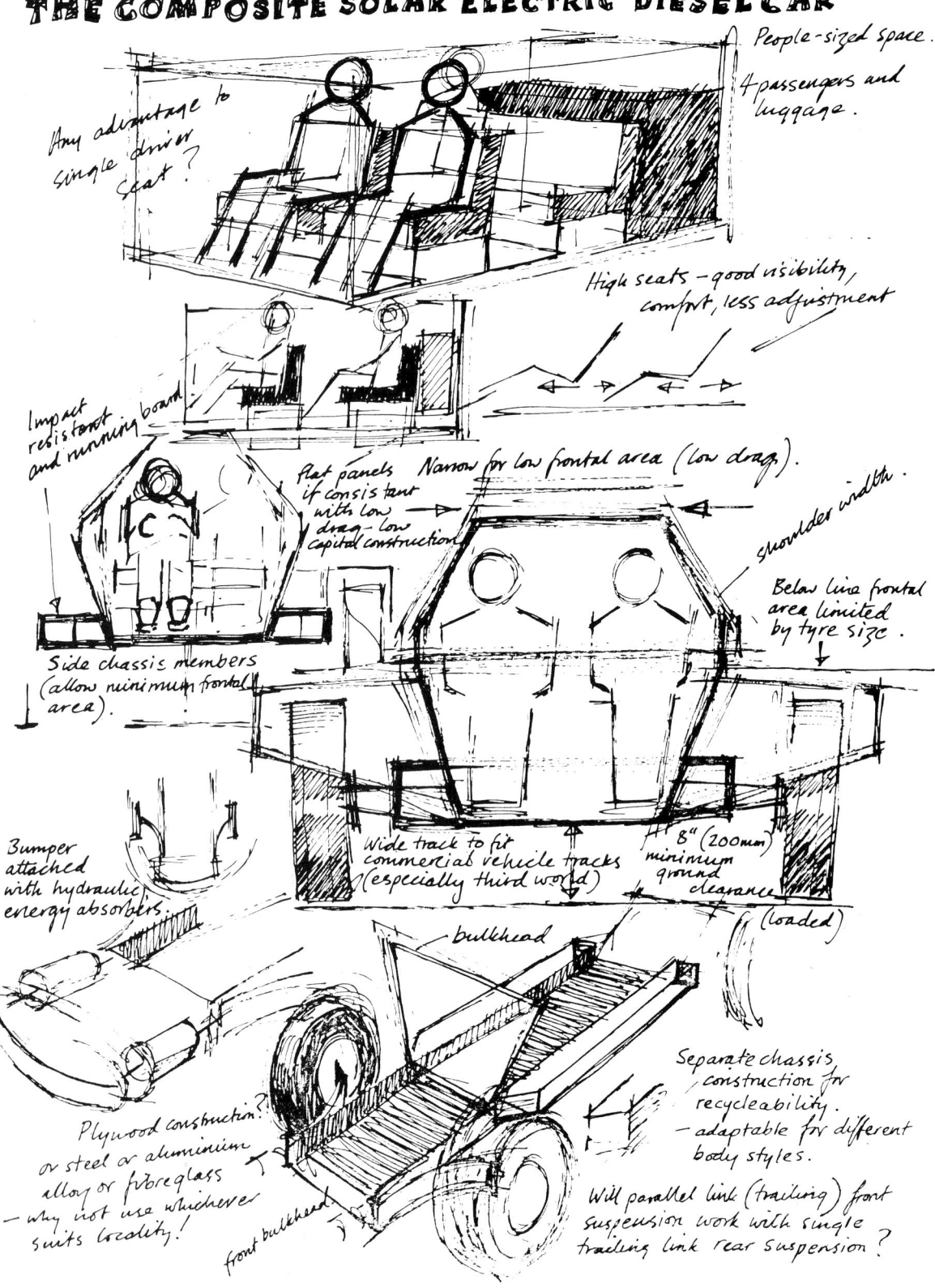

Any advantage to single driver seat?

People-sized space.

4 passengers and luggage.

High seats – good visibility, comfort, less adjustment

Impact resistant and running board

Flat panels if consistent with low drag – low capital construction

Narrow for low frontal area (low drag).

shoulder width.

Below line frontal area limited by tyre size.

Side chassis members (allow minimum frontal area).

Bumper attached with hydraulic energy absorbers.

Wide track to fit commercial vehicle tracks (especially third world)

8" (200mm) minimum ground clearance (loaded)

bulkhead

Plywood construction? or steel or aluminium alloy or fibreglass – why not use whichever suits locality!

front bulkhead

Separate chassis construction for recyclability. – adaptable for different body styles.

Will parallel link (trailing) front suspension work with single trailing link rear suspension?

800-900 lbs. maximum weight. (400 kgs approx.)

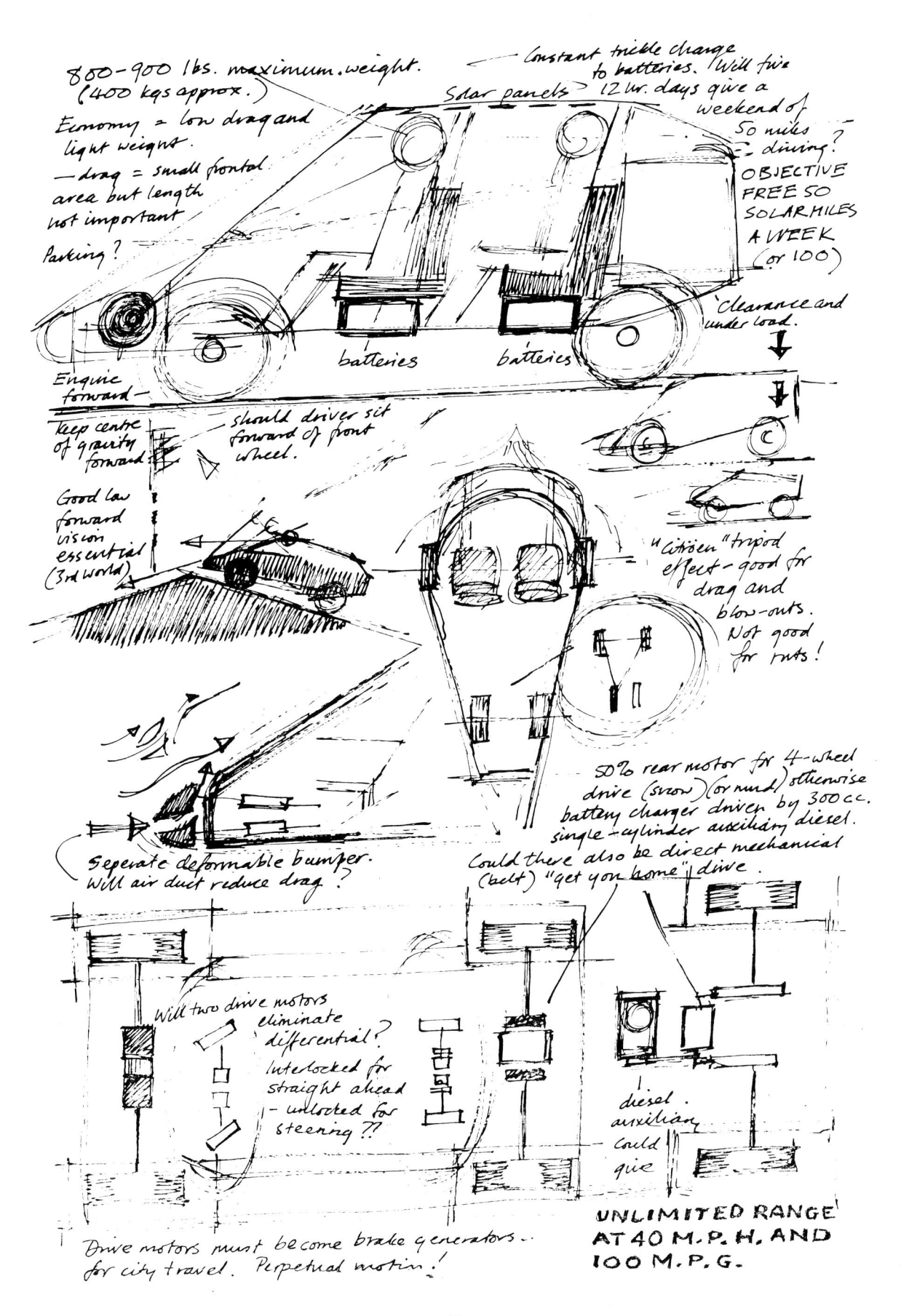

Economy = low drag and light weight.
— drag = small frontal area but length not important
Parking?

Constant trickle charge to batteries. Will five 12 hr. days give a weekend of 50 miles driving?

Solar panels

OBJECTIVE FREE 50 SOLAR MILES A WEEK (or 100)

batteries   batteries

Clearance and under load.

Enquire forward

Keep centre of gravity forward

should driver sit forward of front wheel.

Good low forward vision essential (3rd world)

"Citroën" tripod effect — good for drag and blow-outs. Not good for ruts!

Seperate deformable bumper. Will air duct reduce drag?

50% rear motor for 4-wheel drive (snow) (or mud) otherwise battery charger driven by 300cc. single-cylinder auxiliary diesel.

Could there also be direct mechanical (belt) "get you home" drive.

Will two drive motors eliminate differential? Interlocked for straight ahead — unlocked for steering??

diesel auxiliary could give

UNLIMITED RANGE AT 40 M.P.H. AND 100 M.P.G.

Drive motors must become brake generators for city travel. Perpetual motion!

65

other parts for numerous vehicles, often thought to be the product of other countries. The structure of the motor industry hasn't changed a great deal over the years.

France continued to be the world's main exporter of cars until 1912. But, by 1906, America had taken over as the world's largest vehicle manufacturing country. A year later the Model T Ford was introduced.

Famous for the mass production methods used in its construction, the Model T is equally important in the history of vehicle design. It was, in a way, the world's first, and to date the world's last, real multi-purpose vehicle.

Over 15 million Model Ts were sold between 1907 and 1927. They had a range of bodywork from buggies and sedans to vans, pick-ups, flat beds and even tractors. By 1920 half of all motor vehicles in the world were Model T Fords.

Built as much for the American farmer as for the city dweller, this most famous of all Fords has nearly a foot of ground clearance under its axles, 30 percent more than the average four-wheel drive today.

It was cheap, has a wide track, sensible wheel base and is light. In the standard body form, despite a 3-litre engine, it weighs 650 kilograms. Less than half the weight of typical modern all-terrain vehicles.

The suspension, although not independent, used transverse springs to allow contour following by articulating the axles as well as absorbing vertical wheel movements.

Altogether a superb and seriously practical example of elegant vehicle design, it is one that still lays down much of the essential engineering specifications for an on and off-road vehicle.

In 1927, as American roads improved, Henry Ford abandoned the multi-purpose concept and joined the competitive pack with the stylish but conventional model A. That was the end of Ford as the world's largest car manufacturer and virtually the end of America as a source of innovative vehicle technology.

The initiative had returned to France.

In 1925, a 1923 Citroen 5CV was the first car to be driven around Australia. Fifty years later Jim Reddiex celebrated the anniversary of this achievement by driving a 5CV around Australia again. It averaged over 30 miles per hour and 40 miles per gallon and broke exactly one headlamp bulb – more than could be said for the modern back-up vehicles.

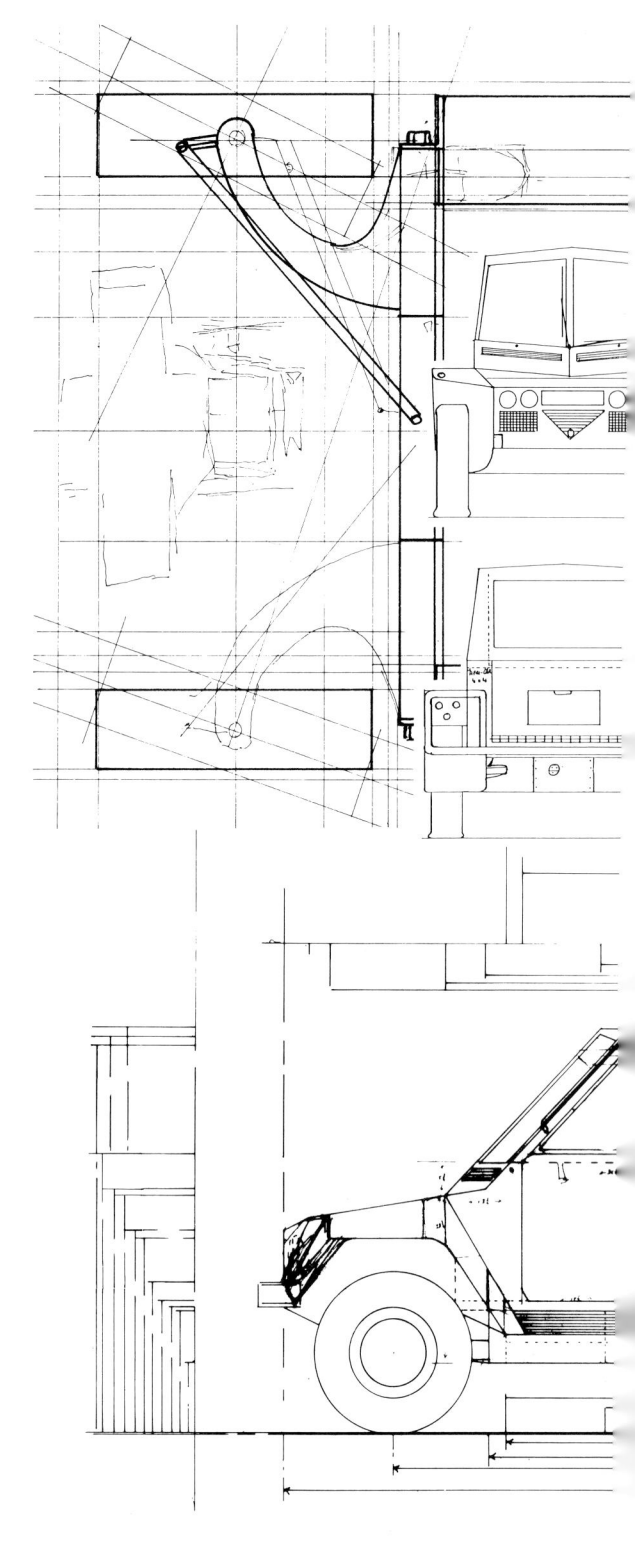

clearance and suspension, giving substantial axle and wheel movements. They were also the first cars in Europe to be sold with electric starters as standard.

Cheap and available, they opened up new markets, particularly with women, making Citroen Europe's largest car manufacturer.

But Citroen espoused the American mass production steel body techniques. The European market, like the Third World market of today, was not yet ready to absorb the numbers of cars necessary to make such production methods viable.

The aftermath of the financial crash of 1929 had bankrupted many European car makers. Citroen was also in trouble but he decided to take one last gamble to save his company. In so doing, he changed the future of the motor car.

A secret think–tank and design group was set up. No one at Citroen knew what was going on. The factory was closed down, completely rebuilt in a few months, and even the production manager had no knowledge of the new model to be built.

When the Traction Avant was announced in 1934 it was a sensation. Few of the ideas taken on their own were completely original, but their combination in a mass production vehicle was unique. It was, and still is, the modern motor car, with front wheel drive, independent front suspension, hydraulic brakes, wide track and low centre of gravity. That's just the start of the list. Plenty of cars of the 1980s still have to catch up with the full specification.

The Traction, already in production in 1934, was the complete antithesis of a vehicle then at an early design stage in Germany.

The Volkswagen Beetle with its rear engine, unstable swing axle rear suspension and inconvenient layout, had only three things going for it. Independent suspension, even if the design was potentially lethal, an engine that was air–cooled and, on the whole, it was a well developed and reliable vehicle.

Conceptually the Beetle was a dead end. It is true that 25 million Beetles were sold as against less than a million Traction Avants, but that only underlines the folly of the consumer which has often shaped the his-

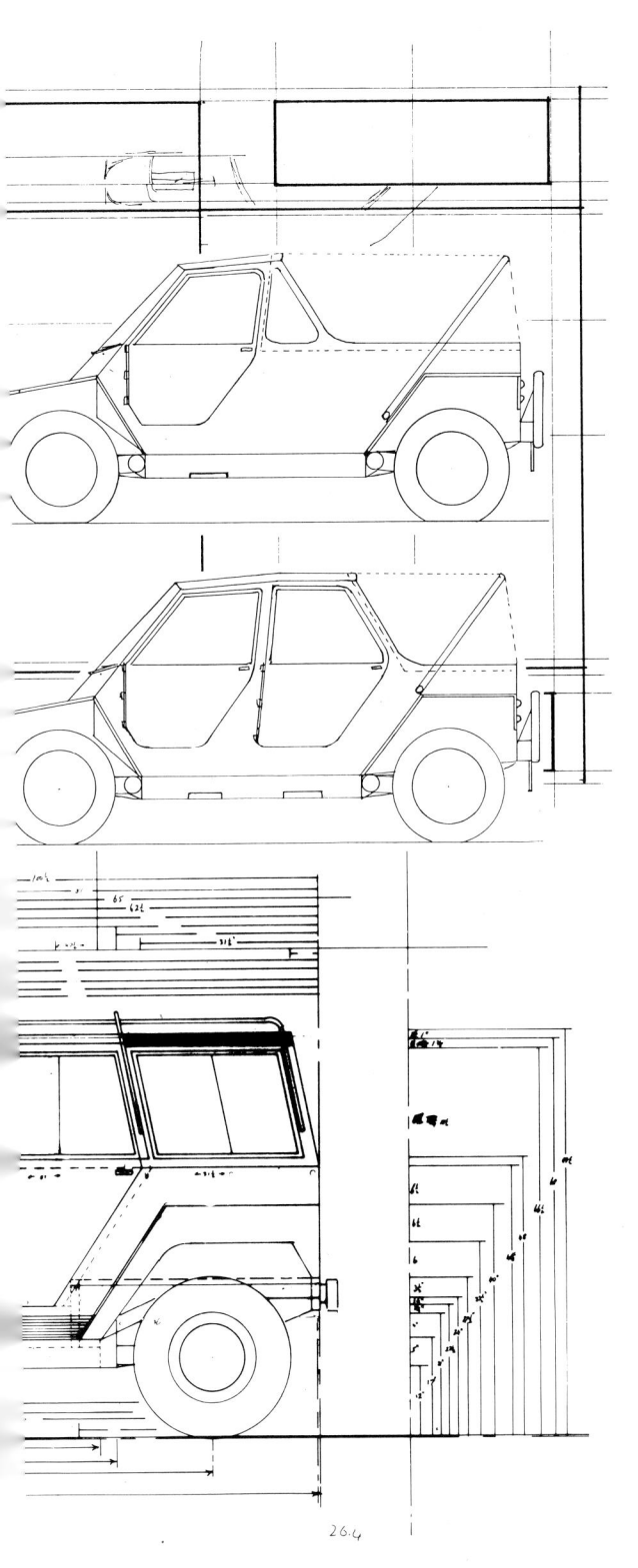

Andre Citroen, inspired by Henry Ford's ideas, built his first cars in Paris in 1919. In post World War I Europe he started with 30,000 confirmed advance orders.

Simple and practical like the Model T, the first Citroens had plenty of ground

tory of the motor car.

Since the mid 1930s, more than 300 million copies of the Citroen concept have been built and sold, as opposed to less than 100 million inspired by the VW Beetle.

Andre Citroen succeeded in getting his new model into production but the company was bankrupt. Citroen died in 1935 and Michelin, the biggest creditor, picked up the pieces and set about rebuilding the shattered company.

In 1955, 20 years after the launch of the Traction came its successor, the DS or Deesse, 'Goddess' as the initials pun in French.

Nothing remotely like it had ever been seen before. It was streamlined, even by modern standards, and had active hydraulic and gas self-levelling, independent suspension with variable ground clearance and proportionally variable positive power braking. Part of a package that left little scope for future car designers.

Since the introduction of the DS, there has been permanent four-wheel drive and anti-locking brakes on the 1959 Jensen FF. For the moment, apart from electronic peripherals and perhaps four-wheel steering, that seems to be the end of the development of the modern motor car.

Five Citroen B2 vehicles made the first motorised crossing of the Sahara in the winter of 1922/23. Fitted with Kegress half-tracks at the back, the Citroen B2s have all the characteristics which are essential to a desert-crossing vehicle today. It's perhaps more exposed than we would like, but in other respects, all the desert applications one would think of today, are on this car.

The old Citroen is a remarkably complete vehicle. Apparently, a few years ago, a battery was attached, the engine was started, and off it went. I think, if they had some new tyres for it and some belts for the half track, it's quite possible that it could head for the Sahara once more.

With a first crossing time of three weeks from Northern Algeria to Timbuktu, the 65 year old veteran could compete with modern overlanders.

The Kegress-tracked Citroens were developed becoming smarter and faster. In the late 1920s and early 1930s they pioneered

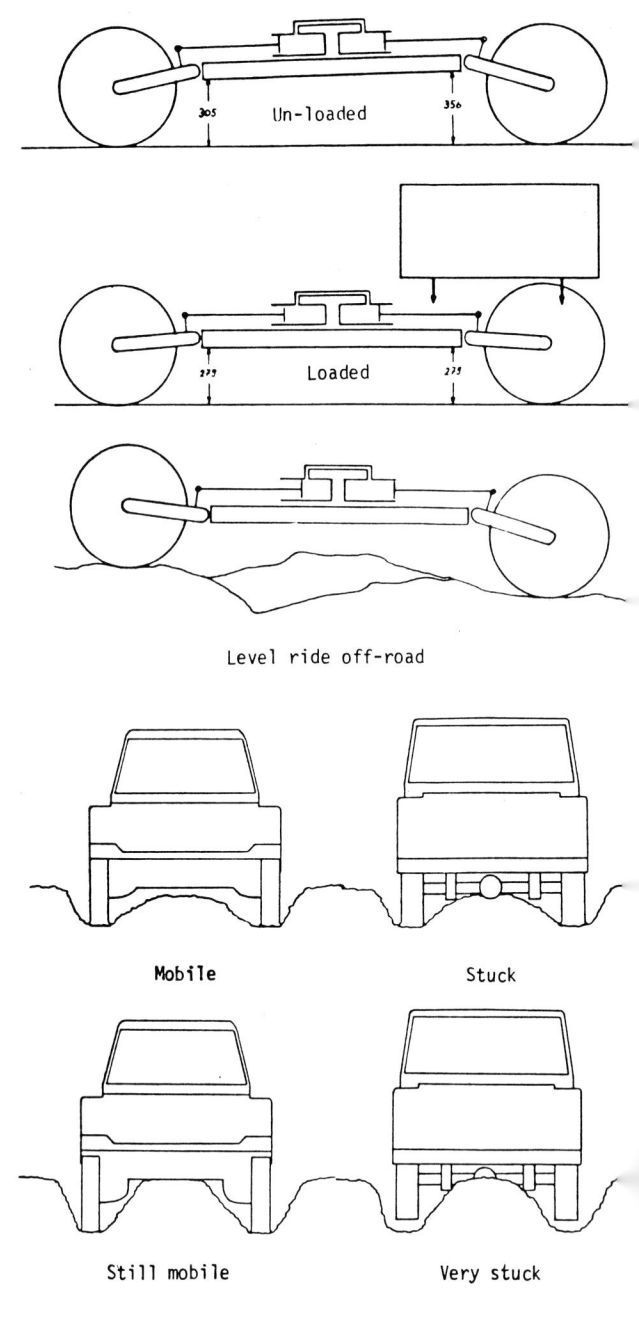

Illustrations showing the dynamic capabilities required for the AFRICAR and how they are achieved

routes throughout Africa and Asia. But they were still slow, uneconomical and cumbersome. With no Model T Ford to fall back on, the world needed something cheaper, more agile and universally useful.

Between 1934 and 1936, under the new management of Michelin, the real joker in

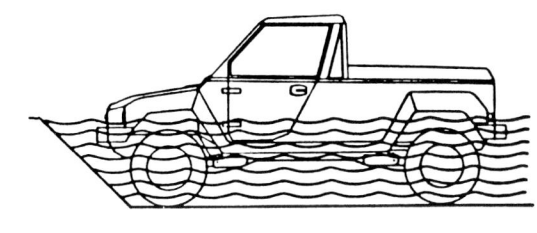

Fording to 750mm

Angle of attack 52 degrees

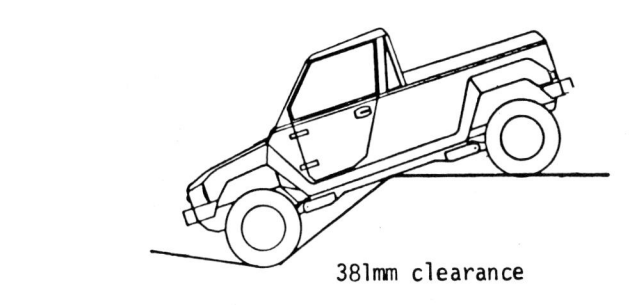

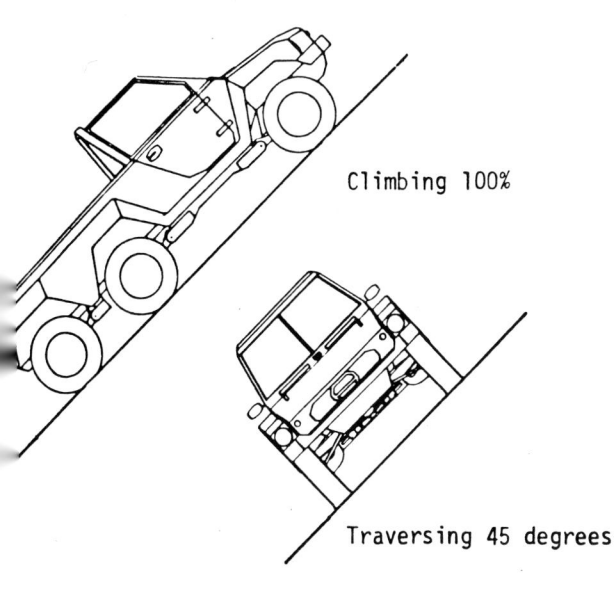

381mm clearance

Climbing 100%

Traversing 45 degrees

the Citroen pack was played.

With a design brief of an umbrella over four wheels, capable of carrying four adults and a basket of eggs over a ploughed field without breaking an egg, the extensive family of 2CV Citroens was born.

Faced with a rigorous specification for the design and use of a car for Africa, the 2CV turned out to be the only useful example of proven technology suitable to function under African conditions, that I could find from the thousands of models developed and marketed in a hundred years of the motor industry.

But the 2CV itself, brilliant as it is, is not really a car for Africa. It's too small, carries too little load, has insufficient ground clearance and suffers from corrosion in its vulnerable body panels.

It is the concept and the uncompromised dynamic solutions of the 2CV that are so valuable as inspiration. As well as the air-cooled engine, front-wheel drive, in-board brakes and all the other remarkable detail.

On the road, especially bad roads, the 2CV's abilities are self evident to the driver. Under the skin it is an engineer's delight. Amongst many ingenious features, it is the suspension that is so exceptional.

Simple leading and trailing links allow large vertical movements of the wheels without the dangerous changes of camber of a VW or a BMW. There is also none of the scrubbing across the road that reduces grip and causes a loss of adhesion.

The only problem with the 2CV set-up is the ground clearance. The chassis is underslung and the springs, operated by pull rods, are positioned low.

The designers of the 2CV solved most of the dynamic problems of vehicle ride and stability, both on and off the road. In adapting their 50 year old solutions to a car for Africa I needed to make some changes and improvements.

I had to lift the chassis and mount the springs above the arm pivots, activating them with push rods.

By mounting the suspension arm pivots at the bottom of the chassis rather than on top of it and then positioning the springs above the pivots, rather than below, I could gain more than four inches or 100 mm of ground clearance. I also sought better damping, better controlled inter-connection and more extreme stiffening of the suspension under load.

The simple lightweight Hydragas unit provided everything I needed. Manufactured by Dunlop, it had been developed over a period of 12 years by Alex Moulton, famous for the original Mini rubber suspension and small–wheeled bicycles, and by Tony Best. Activated by push rods, it could be mounted high for maximum ground clearance with out any complex linkage.

A vehicle designed to run on unpaved roads must obviously have adequate ground clearance to clear the 10 to 12 inch hump between the ruts made by trucks. Conventional four–wheel drive vehicles have about eight to nine inches which is simply not enough.

Equally, it must have a wide enough track to ride in or straddle the ruts. Most four–wheel drive vehicles are too narrow tracked to do this, and become unacceptably dangerous to drive, even on main roads.

With the AFRICAR, for more extreme circumstances, it should be possible to sit on the mud, letting the independently sprung wheels find bottom and provide traction to drag the car free.

For long African distances, and to provide maximum passenger and load capacity, an upright seating position is essential.

But such a wide vehicle would be too heavy and have a huge frontal area leading to high wind resistance and low economy.

To counteract this, the passenger compartment would need to be narrower than the track. So the characteristic shape of the

△
The original wooden mock–up of the AFRICAR, built to test ergonomics, door apertures and visibility

First AFRICAR prototype on its initial day's testing, August, 1982
▽

AFRICAR identified itself.

Using conventional components, such a large ground clearance would lead to a high engine position and poor forward vision. Visibility is everything on bad roads and off the road.

But high ground clearance to an extent demands a high engine position. So, from the beginning, I knew that the AFRICAR would have to have its own specially developed flat, low profile engine and integral transmission to provide a low sloping bonnet.

A decision had already been made to use plywood and epoxy resin for the main chassis structure. A construction method used on racing boats that have to be as light and as strong as possible and have to spend most of their lives in salt water.

Sealing the wood with epoxy meant that we could use soft wood as a renewable resource and there would be little energy consumed in producing the materials or building the vehicle when compared with a pressed steel car. Plywood has the further advantage of needing only low cost machinery to work it precisely.

For prototype work simple hand tools could be used. For production, a computer-controlled profile cutter would undoubt-

△
The second AFRICAR prototype in the wind tunnel at MIRA, Motor Industry Research Association, U.K.

'Bessie' being tested on the edge of Salisbury Plain
▽

The night before leaving for India

On the start line in New Delhi just four months after design started. Drivers are Doug Stewart and Jim Reddiex

Wooden patterns for the third generation prototypes in the workshop at Coalville, Leicestershire

First Himalayan Rally pick–up nearing completion

ably be needed, but at a fraction of the cost of working steel with presses and press tools.

A full size mock–up was built to check the ergonomics of seating position, door openings and visibility.

A strong steel roll cage was used to give passenger protection and to form the door openings and screen surround. Initially welded up with square steel tubes, this frame could evolve into simple rolled section and continuous bent tubes.

A steel sub–frame to hold the engine, bonnet and front wings, was connected via the roll cage and the main suspension sub–frame to the central ply chassis.

The first drive in a completely new prototype vehicle is as nerve–racking as it is exciting. For the AFRICAR life started as it was intended to go on, off the tarmac.

It was a good day. The ride was controlled pretty well regardless of the surface, the upright driving position was comfortable, visibility was good all round and the handling, surprisingly nimble.

Initially the maximum speed was about 70 miles per hour, suggesting that the aerodynamics were about as bad as I had expected. But the promise was there and it was certainly worth going through to the next stage.

Bessie was the second prototype and later became the first AFRICAR to be made street legal.

The shape had been radically altered at the front to try and improve the air flow. But a trip to the MIRA wind tunnel proved otherwise.

The divided and raked–back windscreen, was deliberately designed to give good visibility and to avoid damage from flying rocks and stones spreading right across the screen. It was splitting the air flow to the sides of the car rather than letting it flow smoothly over the roof. A CD factor of .62 was hardly good enough, even for an all–terrain vehicle.

It was, quite literally, back to the drawing board. With the basics of adhesion, ride and steering working well, the ergonomics sorted out, it was essential to improve the drag factor.

A new prototype was built as the pattern for the Arctic Equator vehicles.

With no components in common with the earlier prototypes, this new vehicle, a pick-up, went from the first line on the drawing board to running in the Himalayan Rally in exactly four months.

We fell out of the rally after about a thousand kilometres with fuel pump and vapourisation problems. But while running, and with just 65 horsepower, we made 5th fastest time on one competitive section and consistently held 7th place ahead of a Range Rover, a Volvo, Subarus and others. Evidently we had succeeded in improving the aerodynamics.

As we left Paris to continue our journey to Africa, the Arctic–Equator cars, overloaded and still with only 65 horsepower, ran freely on the autoroute.

We were able to cruise at 70 to 80 mph. The six-wheeler was apparently the most aerodynamic with a maximum of over 85 mph or 135 kph.

Speed was never a fundamental objective

John Fitzpatrick driving the Himalayan pick-up on its first test run

for the AFRICAR, but aerodynamics and economy go hand in hand with the huge distances on the African continent. A reasonable cruising ability is a must for any car for Africa.

At Mulhouse, on the French Swiss border, we visited the French National Motor

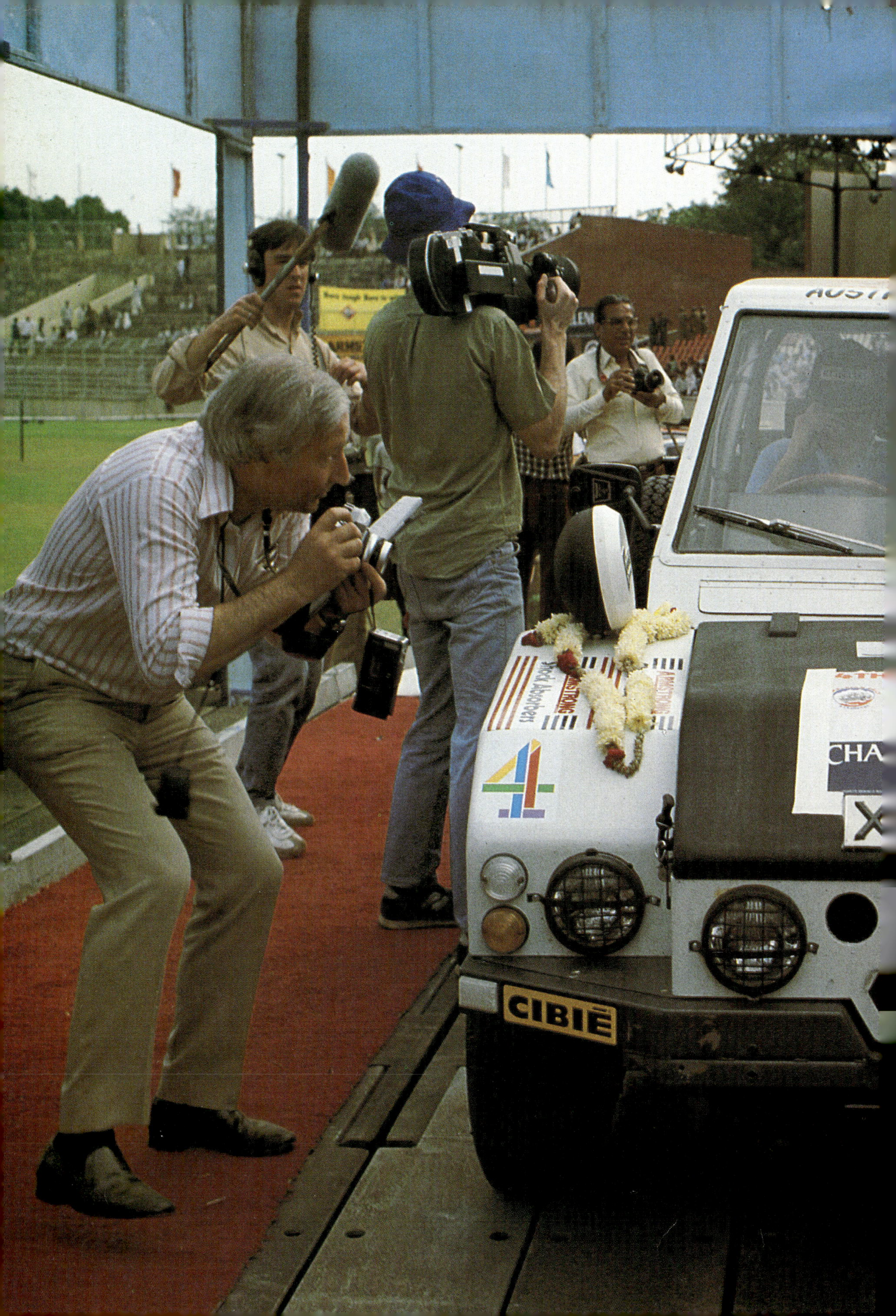
CIBIÉ
ARMSTRONG
Shock Absorbers

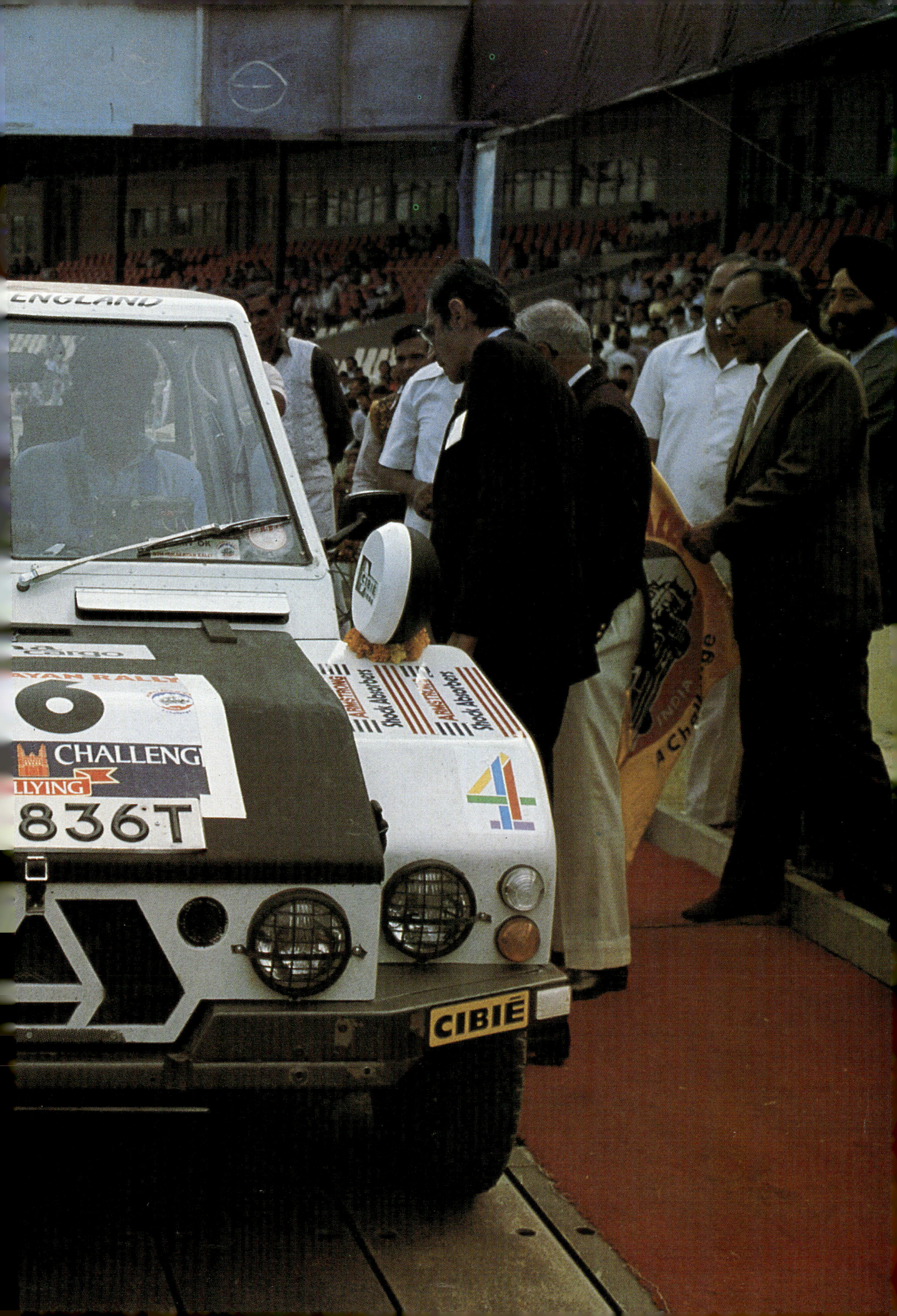
ENGLAND
CHALLENG
LLYING
836T
Shock Absorbers
CIBIÉ

Museum. Based on what was originally a private collection, the museum is not a purpose–built national car archive. Even so we were surprised with what we found.

It was packed with Bugattis, as people would say today, the original yuppie car. All style and elegance and little practical function. Interesting but flawed engineering, often quite old fashioned, even in its day.

In 1984 there was not a Traction Avant, a 2CV or a Citroen DS in sight and few significant Renaults or Peugeots.

Across the border in Switzerland we are on the road to Zermatt. But at Tasch the road is closed to all motor cars. We park

John Fitzpatrick in the workshop at Coalville

Carolyn Hicks trimming the Arctic – Equator cars

Third generation AFRICAR epoxy plywood chassis for the station wagon

alongside everybody else, beside Fords and Fiats, Ferraris and Porsches.

We've stopped here to remind ourselves of priorities. There is a perfectly good road to Zermatt but the internal combustion–engined motor car was banned from the town years ago. Air pollution and traffic congestion were two of the reasons.

Yet Zermatt is a sizeable town and not just an Alpine resort. People walk, there are radio linked horse– drawn taxis, using sledges in winter and wheels in summer, and there are a mass of diminutive electric taxis and utilities.

With plenty of clean hydro–electric power, the electric vehicle is a sensible

The wagon masked for painting

The classic form of the AFRICAR chassis before sub–frames, engine and transmission are added

Charles Best, photographer, working on the pick–up which would become his transportation in Africa

Bugattis at the French National Motor Museum in Mulhouse

A town without automobiles. Zermatt, Switzerland. Only horse–drawn or electric vehicles are allowed

One of the classic Bugatti Royales at Mulhouse

The car park at Tasch where cars must be left before proceeding to Zermatt

alternative in Switzerland.

Almost all towns in the world could operate on a transport mix similar to Zermatt, but modified to be relevant to local energy sources and conditions.

Near Bologna is the factory where Lamborghinis are made. Dream cars originally built out of profits from a post World War II tractor and a central heating boiler empire.

A visit to Lamborghini in the mid 1960s

HOTEL TESTA
HOTEL GARNI TESTA GRIGIA
SPORT TEA-ROOM
HOTEL TESTA GRIGIA
MIETE + TEST VERKAUF SALE + HIRE

Extraordinary roads have to built for the motor car

The Arctic – Equator team at the summit of the Simplon Pass

was one of the major influences in my approach to AFRICAR. For here, contrary to conventional industrial wisdom, the precision parts are actually manufactured in the factory. Either directly by hand or on typical modern computer–controlled batch production machines.

An ideal way of approaching genuine low volume production. Unlike many specialist car manufacturers, Lamborghini make their own engines while buying in the bodies.

The only part that doesn't impress is the intricate welded steel tube space–frame chassis. The heat from every weld distorts the frame, but the Italians are practical when it comes to car building. The frame is beaten and bent, literally with a sledge hammer and a lump of wood into the jig, before the superficial body panels are attached.

Lamborghini is every bit as a much a full car manufacturer as Mitsubishi. The difference is 1,000 vehicles a year compared with 125,000 a year from one plant alone near Osaka. But the processes are quite similar and the economic balance universal. Capital cost of production equipment and labour content per car are equally critical to low and high volume producers.

For the AFRICAR to be produced cheaply in low volume, the balance of labour and capital may be different but the total costs must be the same as in Japan, not those of Lamborghini.

An alternative approach can be found to the south of Japan. In the Philippines the ubiquitous Jeepnee is every bit as stylish as a Lamborghini and is competitive in price with the landed cost of the Japanese equivalent vehicle.

The body panels are cut from galvanised steel sheet by eye with no drawings. Then the voluptuous curved panels are hand made, again by eye, by an army of skilled panel beaters.

Under the skin, Jeepnees use copies of World War II Jeep chassis and more or less any running gear and engines that can be found. Either recycled or imported second-hand Japanese cast–offs.

The AFRICAR epoxy resin and plywood technique requires less skill than the Jeepnee and less expensive imported materials. Its capital requirements are less per vehicle than those of Mitsubishi and it is more precise than the Lamborghini space frames.

No complex jigs and no torture is needed to establish a consistent and exact shape.

△
Descending into Italy from the Alps. It's already spring

On again towards the city that more than 2,000 years ago put the word Roman into a long straight paved road.

At Pompeii we had scheduled a stop to rendevous with my daughter Clare and to check and repack the vehicles. Clare had brought additional supplies for Africa and would take back our discarded Arctic equipment.

Once across the Mediterranean we would be very much on our own with huge distances between points of communication and little possibility of purchasing even fundamentals.

Spare parts for vehicles are scarce in Africa, and at this stage spare parts for AFRICARS are only those that we carry with us.

Back on the road we can feel that we are loaded to the tonne limit on the wagon and the pick–up, and close to two tonnes on the six–wheeler. When we take on extra fuel and the tens of gallons of water needed to cross the desert, we will definitely be overloaded. Not the best way to tackle soft sand but a good test for the cars.

At Trepani in Sicily we board the ferry for Tunis. As Europe recedes behind us I have a sense of relief, 10,000 trouble–free kilometres is a good start.

A project this complex has many beginnings and many watershed points. Without doubt the most difficult aspect has not been the research, design and construction, it has been the question of money – funding.

Two early reports in the late 1970s led in 1981 to what I called the White report, a 70 page document covering all aspects of the AFRICAR concept, including minimum funding levels to develop prototypes.

I showed it to my bankers as a matter of course and courtesy, and then within a matter of a few days of producing it, I was invited to show it to John Charnley, a boat builder from Southampton.

John's response was immediate and positive. We agreed to go ahead on the basis of an equally shared interest and an equally shared financial contribution to the cost of

Belle Isle on Lake Maggiore

Tuscany

◁ Lamborghini body before painting and finishing

Hand beating Jeepnee body panels with no drawings, in Manila, Philippines ▷

Making Lamborghini crankshafts near Bologna
▽

Welding Jeepnee bodies

Italian style

Philippino style

building the first prototype.

He would be responsible for raising the production finance, and marketing. I had provided the initial idea and would look after design and construction. A simple, apparently satisfactory and typical start–up situation.

There was one snag that sat uneasily on the partnership from the start. Charnley had insisted that my estimates for building the first prototype should be drastically reduced.

The first car was up and running within his estimate, but cars are not like boats, they are potential killers and regulated as such. The intended market was the most rigorous in the world. There was, as estimated in the White Report, still much work to be done and much money to be spent.

After eight months, in which he had identified no other source of finance than ourselves, John asked to pull out and I accepted.

Despite the promise of the first vehicle, Carolyn and I were again operating on our slender resources. Needing a revised prototype capable of becoming a road– going version, I worked at home in our garage in Shepherd's Bush in London. But it did look like an early end to the AFRICAR concept.

Then came a summons from our bankers in the West Country. I had never been too happy about Barclays Bank because of their stubbornly maintained involvement in South African.

But we were greeted with much enthusiasm, and our manager volunteered his opinion that this was the sort of project that the bank should be backing. I decided to keep in mind the fact that Barclays is also one of the largest banks in the rest of Africa.

At the Coliseum, Rome
▽

Leaving Rome on the Old Road to Carthage in North Africa (after Carthage: Tunis)

△
In Pompei we re-packed the cars, discarding the Arctic gear and taking the items we needed for Africa

The ferry from Trepani to Tunis
▽

They should at least understand the foreign exchange problems and the need for local manufacture and better vehicles in African countries.

My bank manager offered me the opportunity to go ahead on my own, if we placed all our security, all our borrowing and all our income with his bank.

It was Carolyn's home, her income from editing television news in the trouble-spots around the world and our film-making activities that provided the sole security for the entire venture.

John Woods of Barclays Bank is another person who made a decision that helped make AFRICAR possible.

He also promised that the project would be put forward and recommended to Barclay's Merchant Bank. He said he would become active in seeking finance for the ensuing stages which, as we all knew, would within a couple of years run to six figures and, later, to far more.

△
New AFRICAR meets old African. 403 Peugeot in Central Tunisia

Mr Woods agreed to a staged financing process in modest amounts, to be made available against performance. The intention was for the Arctic – Equator vehicles to be built and the trip to be completed, then the position could be reviewed.

In May 1983, a workshop was set up near Leicester to build the first test vehicle to the new patterns and then the cars for Africa. Work started on the 16th of June.

We moved fast. By October the test vehicle had proved itself and the other three chassis were ready and well on their way to becoming road–going vehicles. In mid–November we made plans to leave for Sweden on the 17th of December and to spend Christmas at the Arctic Circle.

Organisation of the trip, including the critical timing of visas, sourcing of supplies and confirming ferry bookings, had been under way since August.

Along with the building of the cars, costings for the eventuality of full production were being prepared. A major international accounting firm had been briefed to draw up a formal business plan for the future of AFRICAR. All at the bank's request.

While the three new AFRICARS crowded the workshop, discussions were discretely underway for an American franchise and the possibility of funding an initial model production plant in the UK.

Our bank borrowing was at its limit yet still well within the estimated costs. But we had a firm negotiated agreement that we could hold the loan level and continue to use all our income. We calculated that it would give us enough money to complete the cars and the trip.

Then something happened at the bank. We never discovered exactly what.

Income was swallowed, checks were bounced, the plug was pulled. We were told that it was nothing to do with us whatsoever, but we were caught up in something happening at the bank and couldn't avoid the consequences.

We were forced to change our tactics from logic to expediency. The finishing of the vehicles was put off and the departure delayed. At least a year too early and with no possibility of production vehicles in sight, we went public in the media.

It was inappropriate timing, it went against our instincts and any reasonable business practice. But it probably saved AFRICAR.

The response to the press coverage was overwhelming. AFRICAR had really touched a nerve. By January we had had serious enquires from 32 countries, but no means to respond.

The only possibility was to get the cars off to Africa and to prove that they would work under the conditions for which they were designed.

Our bank manager had said that the problem was theirs, not ours, so we acted on his suggestion, we fell back on a temporary expedient – their credit card.

On February 14th, 1984, at six o'clock in the morning with just the time to catch a ferry to Sweden, we finally left the workshop. Our four–wheel drive systems were disconnected as we didn't have the funds to replace incorrectly supplied parts. All our bank accounts, business and personal, were frozen. Our credit cards were overloaded and we barely had enough cash to cover the minimal costs of the trip. And two of the cars had never been on the road before in a completed state.

In Tunisia the sun was really shining and oranges were ripening on the trees. We had a brief first brush with the bureaucracies with which the Third World seeks to defend itself from the exploitive tendancies of the dominant industrial nations.

The wagon and the film equipment had taken four days to get through Tunisian customs. The enforced holiday had been welcome and, probably essential.

For a few days we happily assumed the role of tourists. We followed the Roman roads and the aqueducts, bought fresh vegetables in the abundant markets, peeped into un–affordable luxury hotels and marvelled at the ecological logic of the pit or cave dwellers of Matmata.

Then for the first time on the trip we were confronted by the ordinary road. Ahead lay

◁ **Market in Southern Tunisia**

Evidence of Roman occupation. Signs of the long connection between Tunisia and Europe are everywhere, including a coliseum and the massive aquaduct that used to supply Carthage (Tunis) with water
▽

▲
Our first ordinary road

thousands of kilometres of desert.

Africa started to work its magic. The first day off tarmac the pick–up broke a suspension unit. The problem was not the unit but the chaos caused by financial pressures at the time of departure.

During the last night we had fitted a pair of rear suspension arms without droop or rebound stops by mistake.

There was nothing apart from the weight of the AFRICAR to stop them rotating right round under the chassis.

Africa is unforgiving when it comes to vehicles. Almost any GTI go-cart, flaunting the basic laws of motion, will get away with technological murder on European roads. We had just done 10,000 kilometres without a murmur of a problem, then one desert bump!

It rained in Southern Tunisia. Dried up rivers filled with water and bridges were closed for repairs.

The road had become a slippery sea of mud. It was difficult to distinguish the wet track from the flooded salt flats.

The weather boded ill as we passed the last town in Tunisia and drove on, with only telegraph poles for company, into no–man's land and towards the Algerian frontier and the real Sahara desert.

5
SAHARA

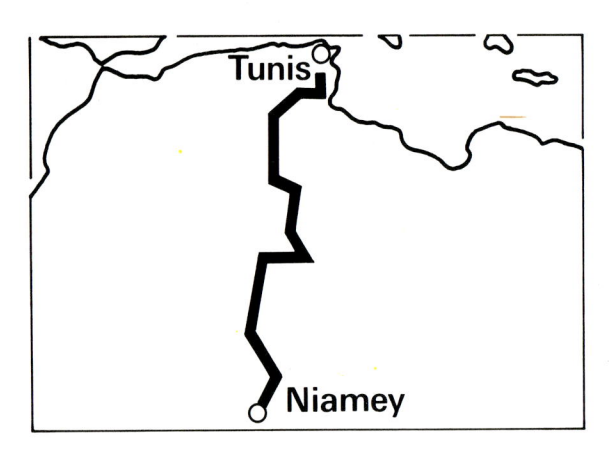

Crossing the Sahara is a little bit like crossing the Atlantic. It doesn't matter how many thousands of others have done it before you, when you set off on a route your life is strictly in your own hands. If you lose

Cars are rarely seen in the Sahara, most travellers use Toyota Land Cruisers or Land Rovers ▷

A Citroen halftrack using the Kegress system from 1930. A sleeker, more modern version, than the one that made the first Saharan crossing ▽

the way slightly and have a mechanical problem you can die – just as easily as somebody in the 1920s or 30s. It's very hard to realise that every time, in a way, is a first time.

The first motorised crossing of the Sahara was quite recent. In the winter of 1922/23 a group of Citroen cars, fitted with Kegress halftracks at the back, completed the desert journey. They took about three weeks, which makes an interesting comparison with modern travellers who rarely take less than two weeks to follow the same route. Many modern travellers take far longer because few escape without major damage to their glossy, contemporary production cars.

We also had our difficulties in the desert. We broke trackrods and suspension pivot pins and we had anticipated failures of the small temporary gearboxes we were using.

But the element of the AFRICAR that might have been expected to fail, the plywood and resin chassis and the lightweight subframes attached to it, gave us no trouble at all. This was in contrast to the many production vehicles we came across that had broken their steel chassis and subframe assemblies.

The mechanical problems we came up against in the desert vindicated my decision to test under conditions of real environmental pressure. Under artificial test conditions the vehicle would have been stopped and repaired.

As designers and would–be constructors we had put ourselves in the position of being customers. We simply had to get to the next oasis and water regardless of the consequences of a broken component.

In the Sahara, without water for just twenty four hours, we would have died.

Our first real Saharan camp ▷▷

Waiting in no–man's land between Tunisia and Algeria ▷

Sealed film equipment at the Algerian border ▽

'Europe and Tunisia are behind us and we are poised on the Algerian border before crossing the Sahara Desert.

The planned Saharan route is via El Oued, Illizi and Djanet to Tamanrasset before heading south across the Republic of Niger to Kano in Nigeria.

But here the problems of AFRICAR shift from the Western casino money market to Third World bureaucrats. Entry with the cars and the film equipment has been refused at the Algerian border. With no visas to return to Tunisia and no suitable documentation to go forward, we are stranded in no–man's land.

Tony Hughes, whose job it is to smooth our way across frontiers, grooms himself for another assault on the isolated border post. The rest of us wait, a little apprehensively, under the watchful eyes of Algerian soldiers.

I have to prepare a new detailed inventory of absolutely everything we possess, in French and in triplicate.

TONY: 'The position is that there is no telephone or other communication equipment at the border post but it is possible that one or two of the group could take a taxi 60 kilometres into El Oued to try and get permission.'

We had attempted to follow the correct procedure for obtaining permits over a period of six months. It is now Wednesday the 28th of March 1984. Our first contact with the Algerian Embassy in London was in September 1983.

In October we contacted the Ministry of Information in Algiers, which, by November had agreed in principle to our project. Then they passed us on in the same month to the Ministry of the Interior, where the buck seems to have stopped, despite repeated representations by the British Embassy in Algiers and directly by ourselves.

During the last three days, Tony Hughes and I have made a 1,400 kilometre round trip to Algiers in the four–wheeled wagon to present our case.

Once again the Ministry of Information had no objection but the Ministry of the Interior were immovable.

So here we are, sitting amongst a pile of sealed film equipment, on the edge of the most famous, most varied and arguably the most beautiful desert in the world – with the prospect of enjoying it for ourselves but unable to share it with others.

But bureaucracy works in subtle ways. The man from the Ministry had finished by saying:'Why don't you just go away and do

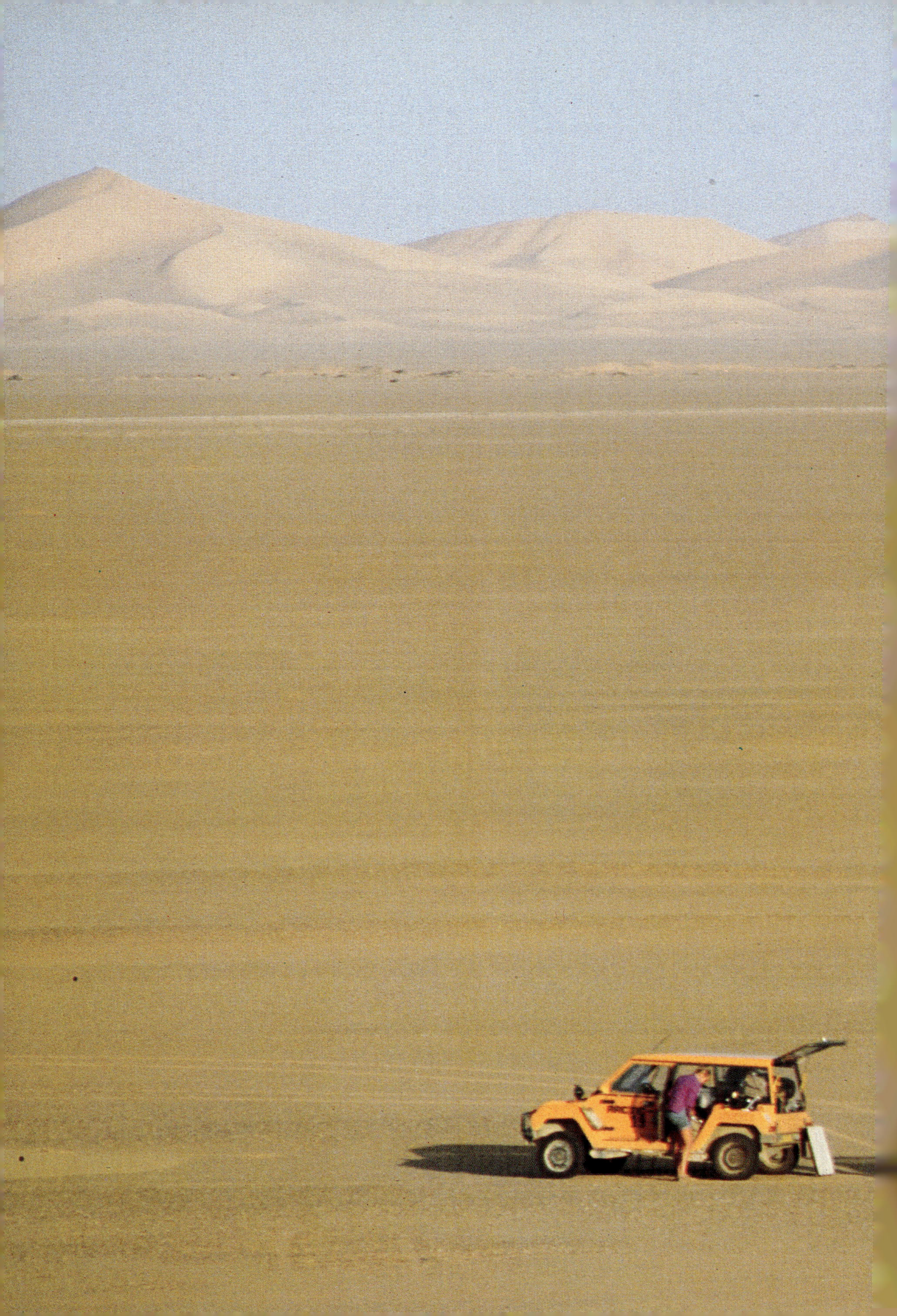

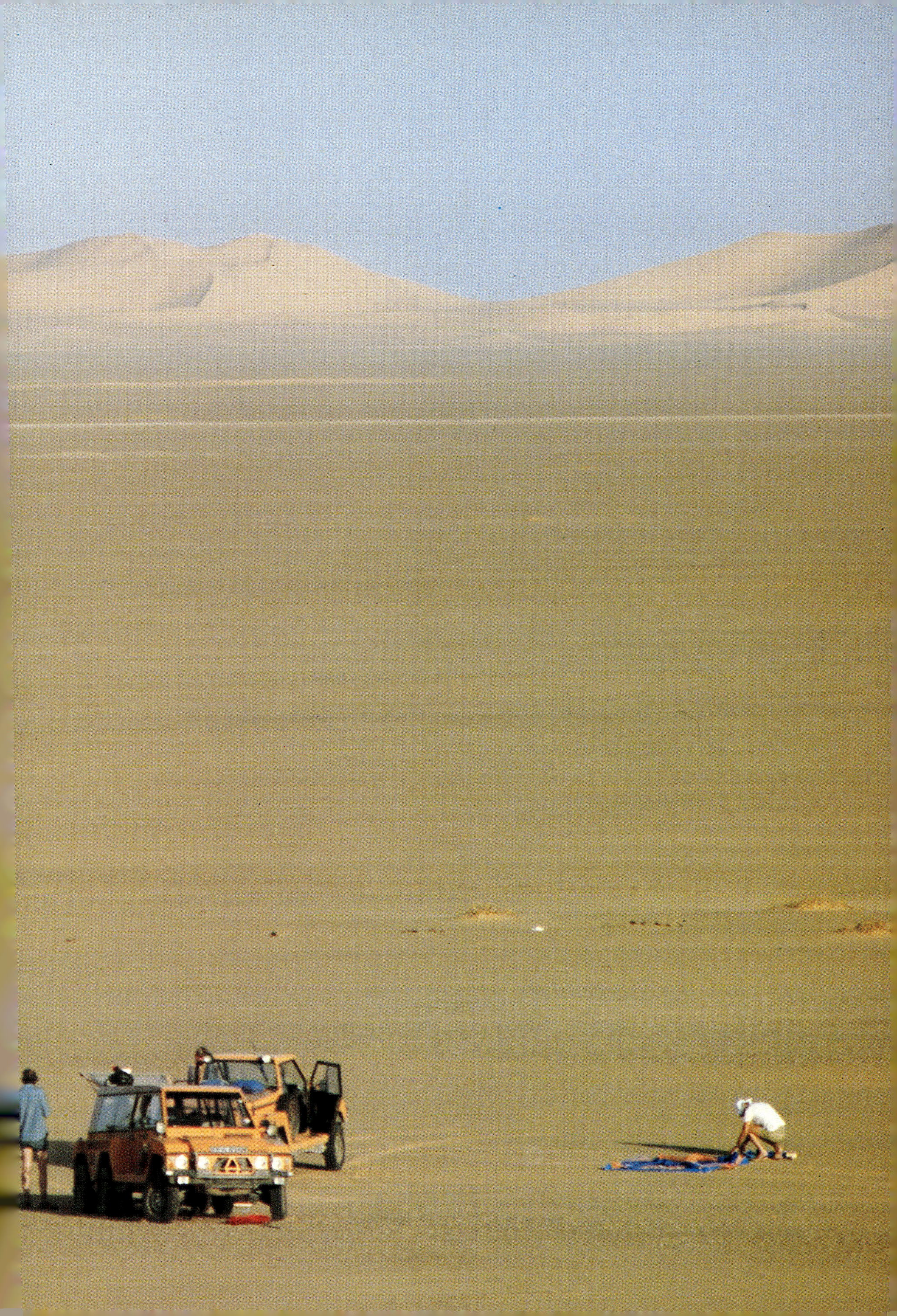

△
The Sahara is always changing. After flat sand and dunes there are mountains ahead

The good tarmac of Northern Algeria rapidly disintegrates as you travel south to the desert
▽

what you came here to do,' leaving unsaid the bit about not getting caught.

So, on the strength of the space between words, we are off south, to find our first real Saharan camp. Charles Best, our photographer, has long anticipated the Saharan sunsets and sunrises as we shivered our way across the February landscape of Europe.

On the northern edge of the Sahara the nights are bitterly cold and dawn is little better – until the sun breaks through and temperature soars.

For part of our second day in the Sahara we ran the desert, keeping station with the good tarmac road which serves the new oil towns of Eastern Algeria. We drove fast in tight formation with our gas springs and independent suspension soaking up even the unexpected undulations of the desert.

After a month and a half on the road each member of the team has settled down to their own specific job. Bob Williams, the driver and mechanic of the six–wheeler, is our first casualty. He had originally asked to come with us to Africa but he did not like the desert. Less than a thousand kilometres into Algeria he collapsed.

We stopped and made sure it was nothing more than sun, heat and perhaps home sickness.

They say disasters come in threes. The second came immediately Liza drove the pick–up to join the tarmac. The bonnet, which had not been fastened correctly, flew up and cracked the windscreen.

The wood and resin bonnet absorbed the impact without damage, only the stay became detached. Driving the pick–up against the Saharan sun was going to be difficult. At least the central screen support had localised the cracks mostly to the passenger side.

It would be 5,000 kilometres before we could replace the screen in Niger.

We have now lost four days at the Tunisian customs, 10 days getting into Algeria and the best part of a day because of Bob's illness. Altogether 15 critical days in relation to our limited resources.

With no reserve money left in England, a broken windscreen and one essential crew member certain to ship home before long, I know that it could become impossible for us to complete the journey to Kenya.

We are running towards Libya and the road is starting to deteriorate. Summer day temperatures of 50 degrees centigrade or more and nights close to freezing, poor foundations and relentless heavy truck traffic is no way to treat a tarmac road.

A trans–Saharan railroad might have been a cheaper and better solution a century ago.

The road takes it's toll. Broken tyres, huge lumps of alien black rubber littered the desert, mile after mile.

It had to happen, the third disaster. I was driving when we hit a big pothole, almost

△
The pick–up's windscreen broken because the bonnet wasn't fixed down early in the Sahara

Repairing the six–wheeler after three wheel rims were bent by a tarmac pothole
▽

six inches deep with hard tarmac edges. We bent three wheels on one side of the six-wheeler but in doing so tested another essential design feature.

We deliberately use lightweight, soft and malleable wheel rims and as a result the impact has been absorbed in the rim without destroying the tyres or distorting the wheel centres.

There is a relentless wind from the west, not gusting, but blowing steadily at about 30 miles an hour. Sliding windows were chosen for the AFRICARS to give maximum ventilation and protection. Cheaper conventional wind-down windows are quite useless under these conditions.

The insulating wood and resin roof protects us from the worst of the overhead sun. But it is still hot – really hot. I am grateful that the steering wheels are not plastic, yet even the leather cannot breath fast enough.

In the heat water is starting to become an obsession.

At Ohanet, besides an experimental farm, we revel in the only pressure water supply that we found in the first 2,000 kilometres of desert.

A nearby source of water was really critical to a group of Swiss bikers. Riding fashionable dirt bikes, they were already plagued with problems such as sand in the chains, overheating and complete loss of compression.

In the desert where water, food, spares, clothes and extra fuel have to be carried, the bikes cannot cope. A dream of riding the dunes stimulated by an advertising image – is shattered by reality.

We have reached the edge of Tassili, a high plateau eroded by wind and ancient rivers. In caves there are thousands of rock paintings depicting a once populated area with water, vegetation and abundant wild life.

The wind funnelling through the eroded valleys has created localised dunes, awesome in size and colour.

Just outside the small town of Illizi is a sign which says, 'This track is dangerous for 200 kilometres'. What it fails to point out is that, for vehicles, it is the surface of the tracks itself that is the danger, not the mountain passes and tight bends.

This road is the main reason we have

100

The Swiss bikers seek the only shade they can find in the desert

Filling all our water cans at Illizi before attempting the Illizi – Djanet road

The magnificent Tassili Plateau broken up by erosion

The danger sign outside Illizi means bad road surface, as much as other hazards

come to Algeria. The next 400 kilometres to Djanet, a small oasis on the Libyan border, is rated in some international overland guides as: 'The worst road in the world'.

As we start to climb towards the top of the Tassili plateau, the AFRICARS seem to be taking it all in their stride, but I remember a chance meeting with two Canadians in a brand new Mercedes four–wheel drive.

Do not go over 15 kilometres per hour, they warned, or you will break everything. It's corrugated rock, not sand.

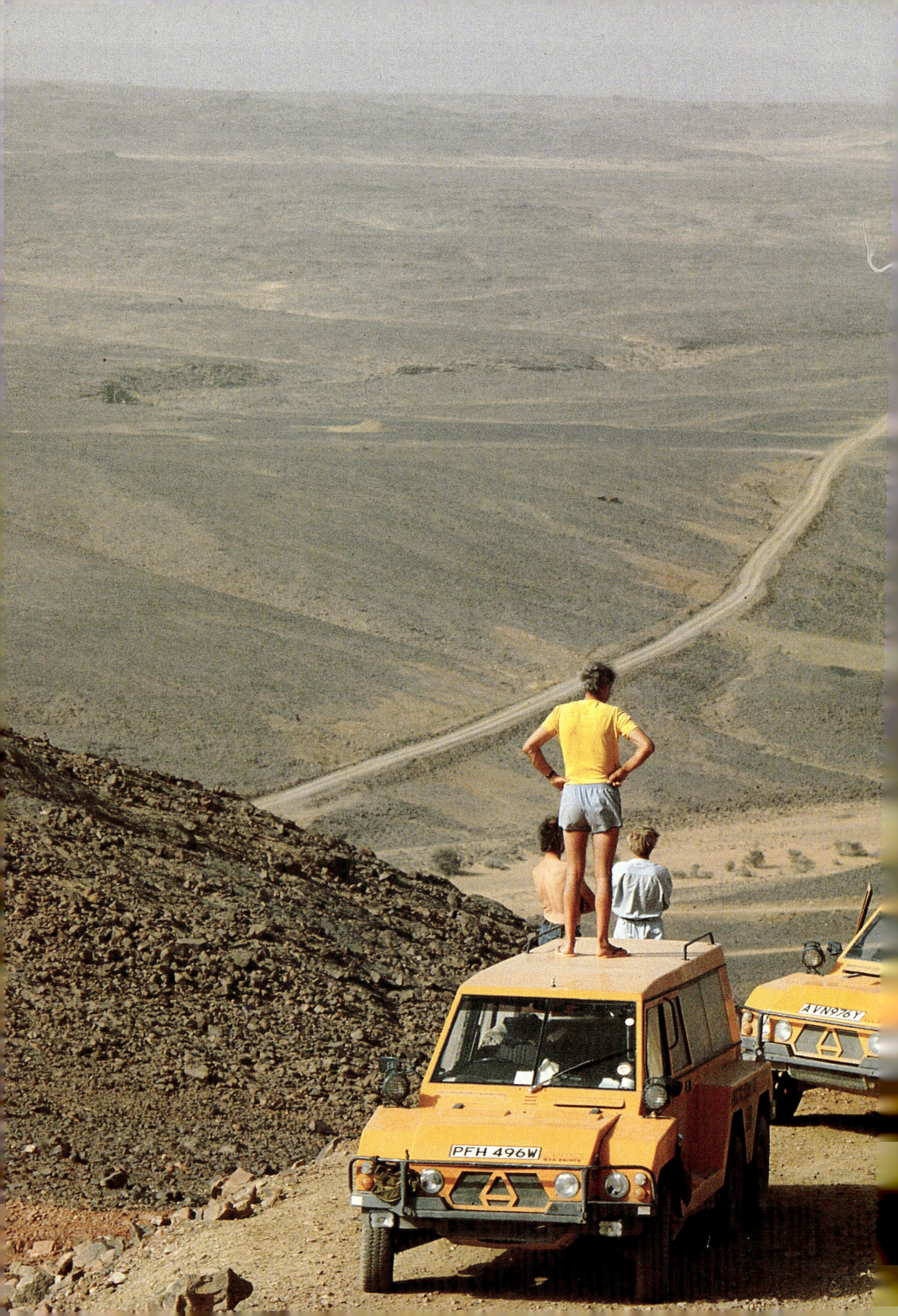
PFH 496W
AVN976Y

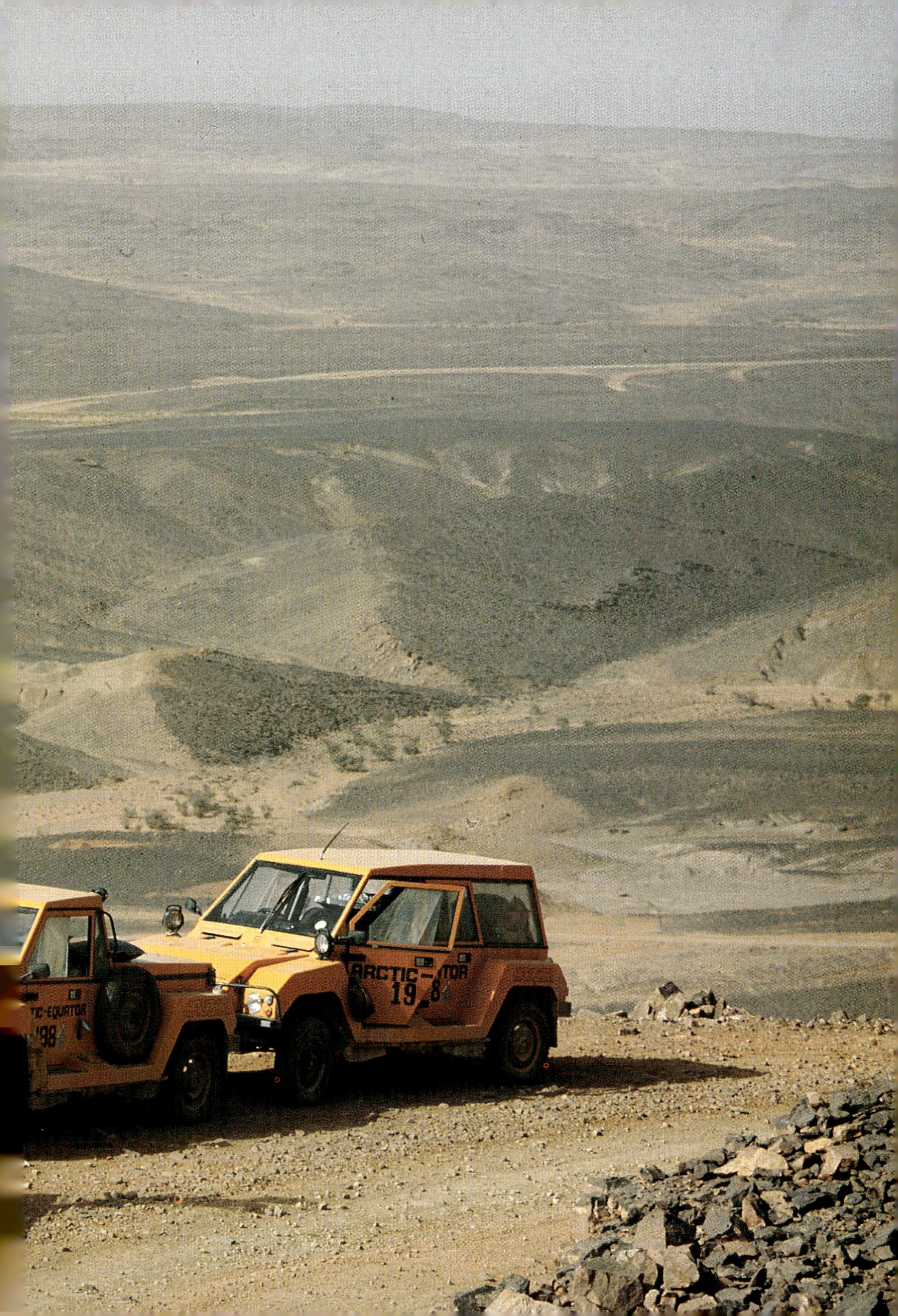
ARCTIC-

Camels do not need a road – ordinary or extraordinary

We're crossing a high plateau covered in rocks and broken up limestone but most of the rock that's left behind is like glass. The vehicles can handle the rocks, it's the corrugations that are really the problem. They are absolutely regular and get you as soon as you drive off the rock on to the sand.

Corrugations are much worse than the rocks because the speed you travel doesn't matter, they will pick up the frequency of the car. Once they've got hold of you you can come right down to three or four kilometres an hour and the car will still be bouncing on them.

We were recommended not to travel faster than 20 kilometres an hour on this road and a number of people suggested 15.

We find with the AFRICARS we are able to travel between 20 and 30 kilometres per hour with reasonable comfort. It's going to take one or two days to do this 200 kilometres unless we want to go up to a speed that could break these cars in half – as it would with any other car.

Danger has always attracted travellers

Corrugations

Climbing further into the Tassili ▷

Overlanders on the Illizi – Djanet road

and every year a trickle of overlanders make their way to Djanet. Mostly they drive Land Cruisers or Land Rovers. The heavy beam axles fly uncontrollably on the corrugations, battering and breaking the vehicles and their contents.

Normal four–wheel drive pick–ups are not up to it, especially over–bodied and over–loaded like the so–called World Cruiser. We found it less than 500 kilometres further on broken, quite literally, in half.

Tassili is about half the area of Spain and the low speeds possible on the tracks make it appear even larger. The climb seems endless, rising by stages from one level to the next.

Each intermediate plateau has a carpet of rocks or stones, each sized precisely as though they have been graded through a giant sieve.

Underneath the rocks there is sand and jagged edges of lava, a surface that competes even with Central African roads for violent steering kick–back on conventional utility vehicles. This has to be negotiated at a snail's pace if the axles, chassis and springs are to survive.

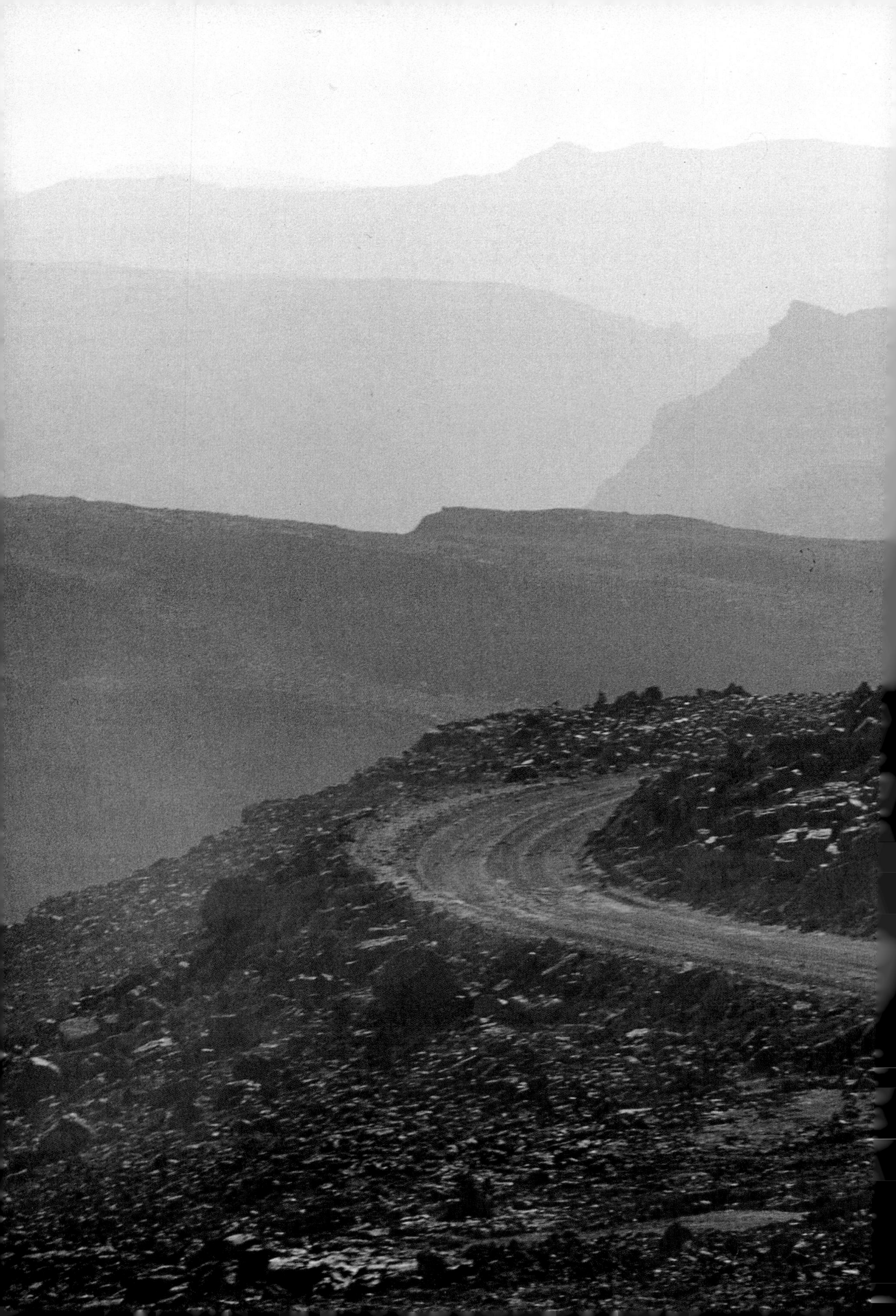

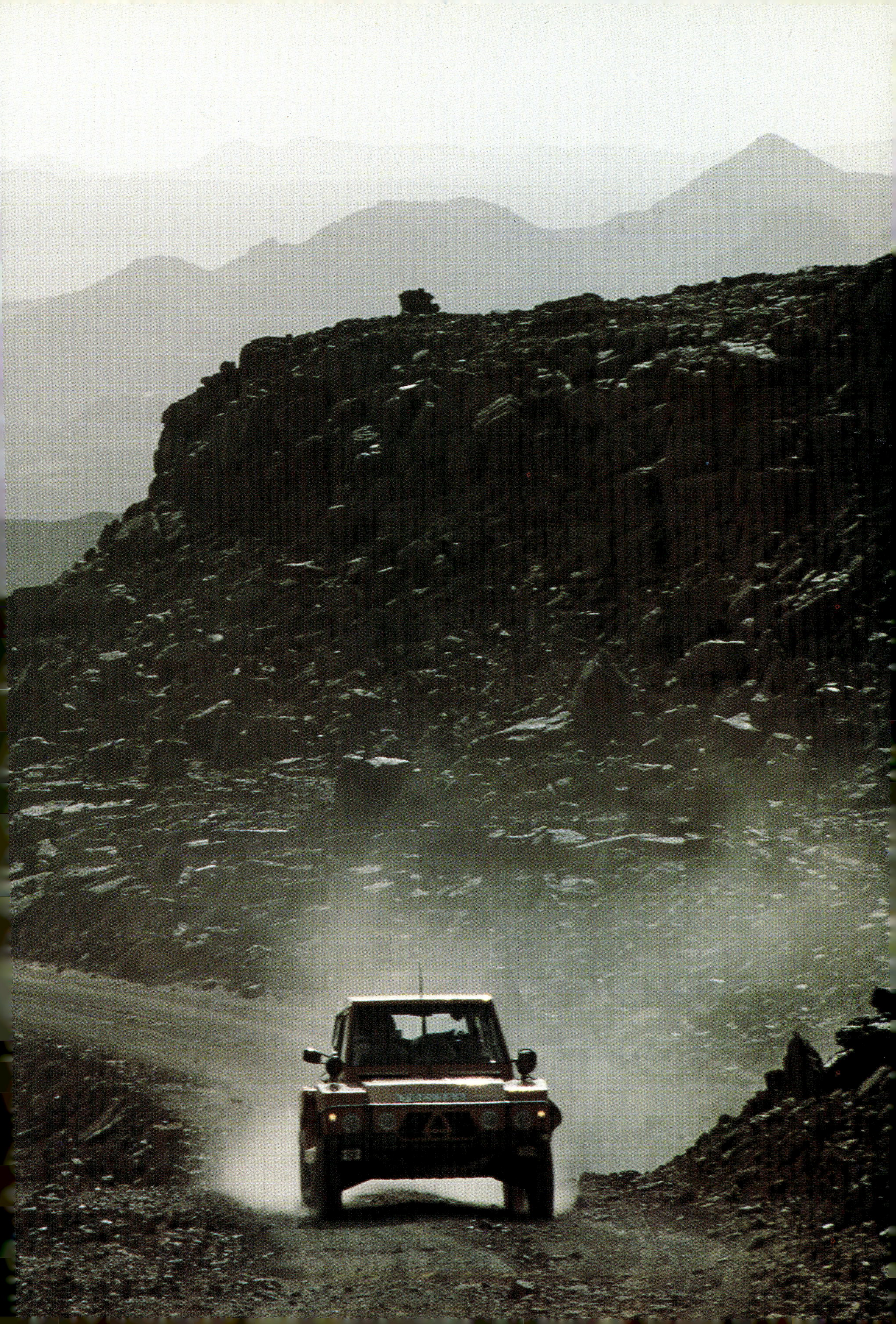

△
Naturally graded rock, carpets the flat parts of the plateau

On a particularly bad day on the piste we met one of the world's great, but generally misunderstood, overland vehicles. It is the independent suspension and low unsprung weight, not the rear engine that has made the Volkswagen bus or combi so effective. The rear engine is in fact the worst feature of this 1930s derived design but air cooling has helped make the legend.

At the highest point of the plateau is a view point which is a meeting place for overlanders. We found the Swiss bikers there, anxious about the route ahead.

Then suddenly it was crowded. A French Toyota and a Norwegian Land Rover. The French group were on a round trip through Djanet and back to France. The two Norwegians were more adventurous, heading for Dakar in Senegal and then on to South America.

The Sahara is full of surprises. Below the Tassili summit is a small volcanic crater lake which holds water all year round and possibly has done so for millions of years. The water is inviting but disappointingly brackish and undrinkable.

Then we found our first real soft sand, just a patch to catch the inattentive and inexperienced. No problem worse than a quick push out.

At the base of the plateau we are back on sand for the last two hundred kilometres to Djanet, but we have to find tracks to guide us, at least to the firmer parts of the piste.

Driving on soft sand requires a learned technique. We have already discovered that letting the tyre pressures down helps, so we know that in future we should use slightly larger tyres.

We have found it is mainly a matter of speed and judgment of when and when not to accelerate. Although we are still running just in two–wheel drive we have, to our surprise, no real disadvantage – even with one ton loads on the four–wheelers and two tons on the six–wheeler.

When we are forced to stop off the hard sand, or when we simply get it wrong,

First meeting with Thomas's Toyota and the Norwegian Land Rover

front-wheel drive allows us to power out with the front wheels riding up on the sand and not digging in.

I now have no doubt that, contrary to normal practice, the AFRICARS should have dominant front-wheel drive built into their all-wheel drive system.

While driving the prototypes on the Illizi to Djanet road, we have experienced another unexpected justification for having the principle drive at the front.

So far we haven't broken any chassis or major components but we have broken track rods and with no spares we have run out of resourcefulness to repair them.

The pick-up is now running on soft sand, with one lazy front wheel and only one wheel steering. That could not be done with rear-wheel drive or with equally split four-wheel drive.

The problem with the track rods vindicated my decision to test under conditions of real environmental pressure. Under artificial test conditions the vehicle would have been stopped and repaired.

The last part of the piste to Djanet was fast and satisfying but Bob had a desperately hot and dusty drive nursing the pick-up along.

Djanet means 'paradise' but for most travellers it means a trip to the little garage on the outskirts of the town.

With the loan of welding equipment we were able to reinforce our track rods so effectively that they gave us no further trouble.

The Swiss bikers had broken carrier and fuel tank mounts. A Toyota Land Cruiser had run all its bearings twice.

The Norwegians had serious problems. They had broken the front of their chassis clean through and they weren't even half way across the Sahara, let alone in South America.

While Bob fixed the track rods we were joined at our camp by two Germans, Bernt and Thomas, travelling with a Land Cruiser. They offered to accompany us to

Our first soft sand

A rear–wheel drive bus inches its way through soft sand before Djanet

Tamanrasset where Thomas would join us as mechanic and Bob could go back to England.

With the steering strengthened we set off at speed, intending to make up for lost time. We were driving west along a flat wide sandy valley with temperatures of over forty degrees centigrade and at speeds of up to 100 kilometres an hour – we overdid it.

For three years we had scoured the world for an efficient lightweight four–wheel drive gearbox. One that would allow us to retain both a low centre of gravity and our ground clearance.

We found that again we were addressing a problem not even considered by the motor industry and we came up empty handed. We would eventually have to design and build our own transmission system from scratch.

Meanwhile a stop–gap had to be found quickly. I chose a small Citroen box which had the right main ratios and approximately the correct layout.

These gearboxes started life almost 40 years ago transmitting less than a third of the power we needed. So we had set off for Africa expecting to rebuild the boxes every 10,000 kilometres or so.

Within a matter of ten metres from one another, two gearboxes failed. The six–wheeler's box was pretty well smashed, but Carolyn's car, the wagon, had only run a bearing and remarkably, once repaired never gave us a problem again.

Despite a spirited race between Thomas and Bob, it took a day to get both the wagon and the six–wheeler back on the road.

The route we are on continues due west to meet the main trans–Saharan highway. We want to turn off and pick up a trail to the south.

The Norwegians have caught up with us and we have all spent half a day looking for the marker, a piece of tin on top of a small stone cairn in a million square miles of emptiness. Amongst the graffiti it bears the all important arrow and legend – TAM for Tamanrasset.

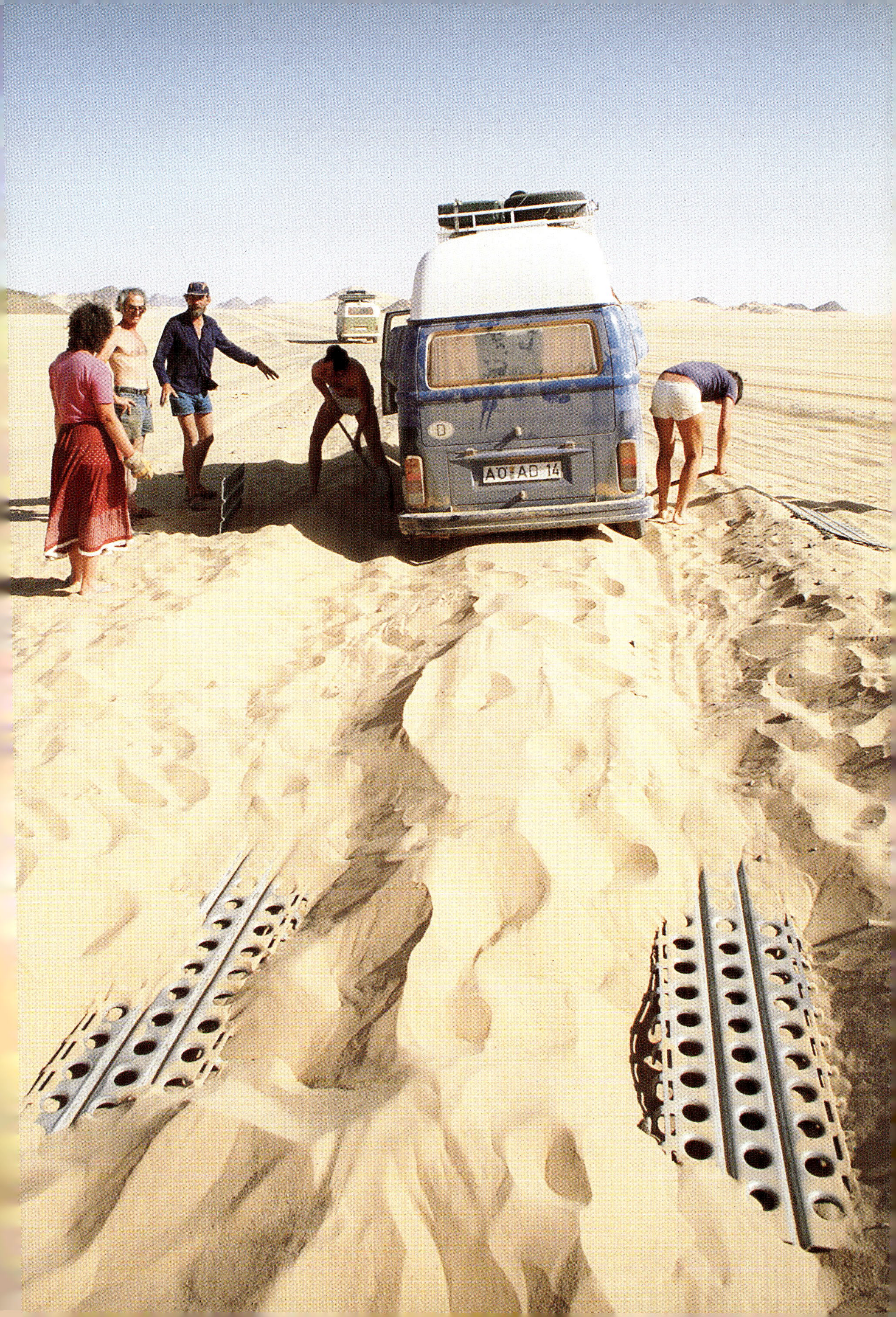
D
AÖ AD 14

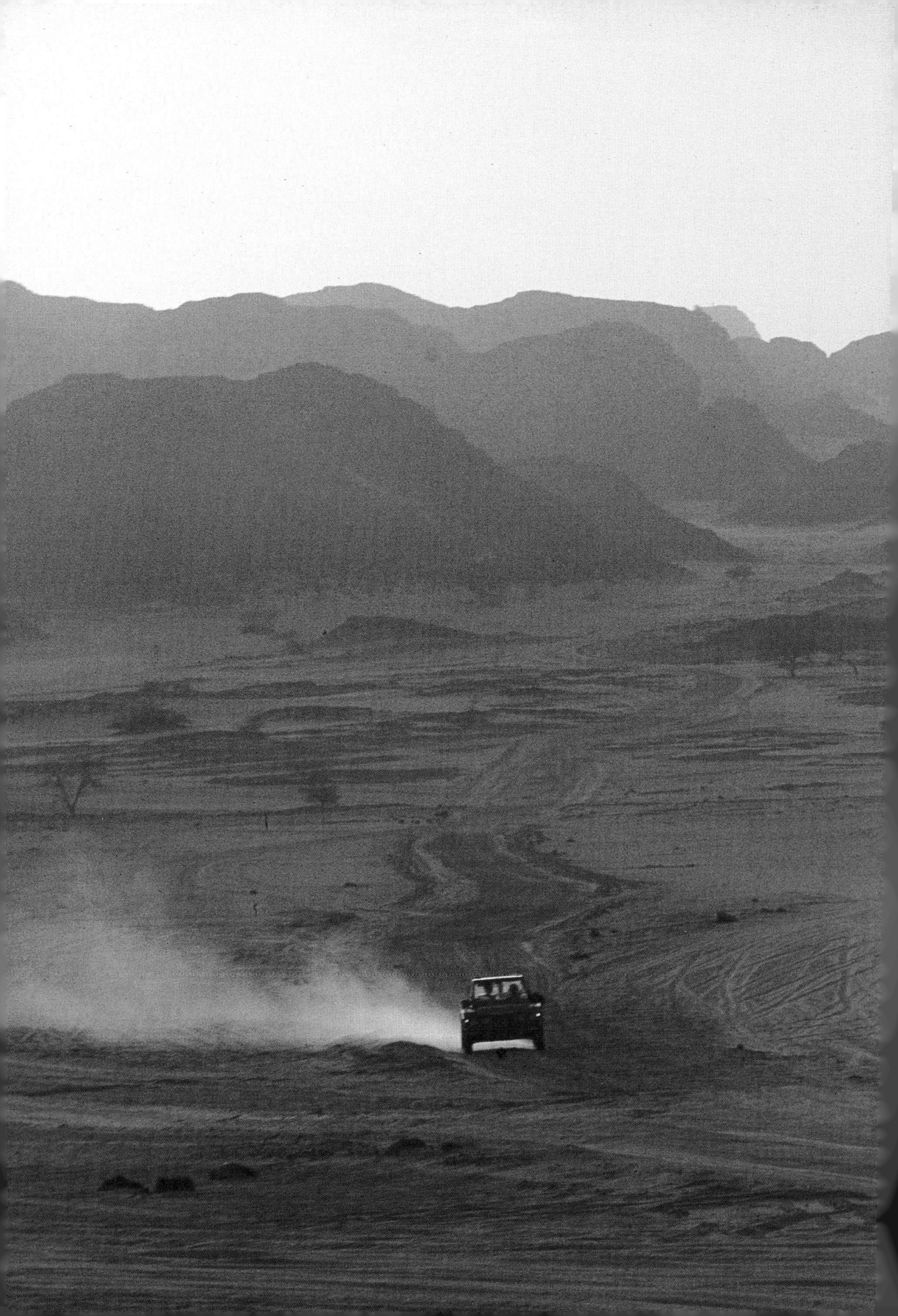

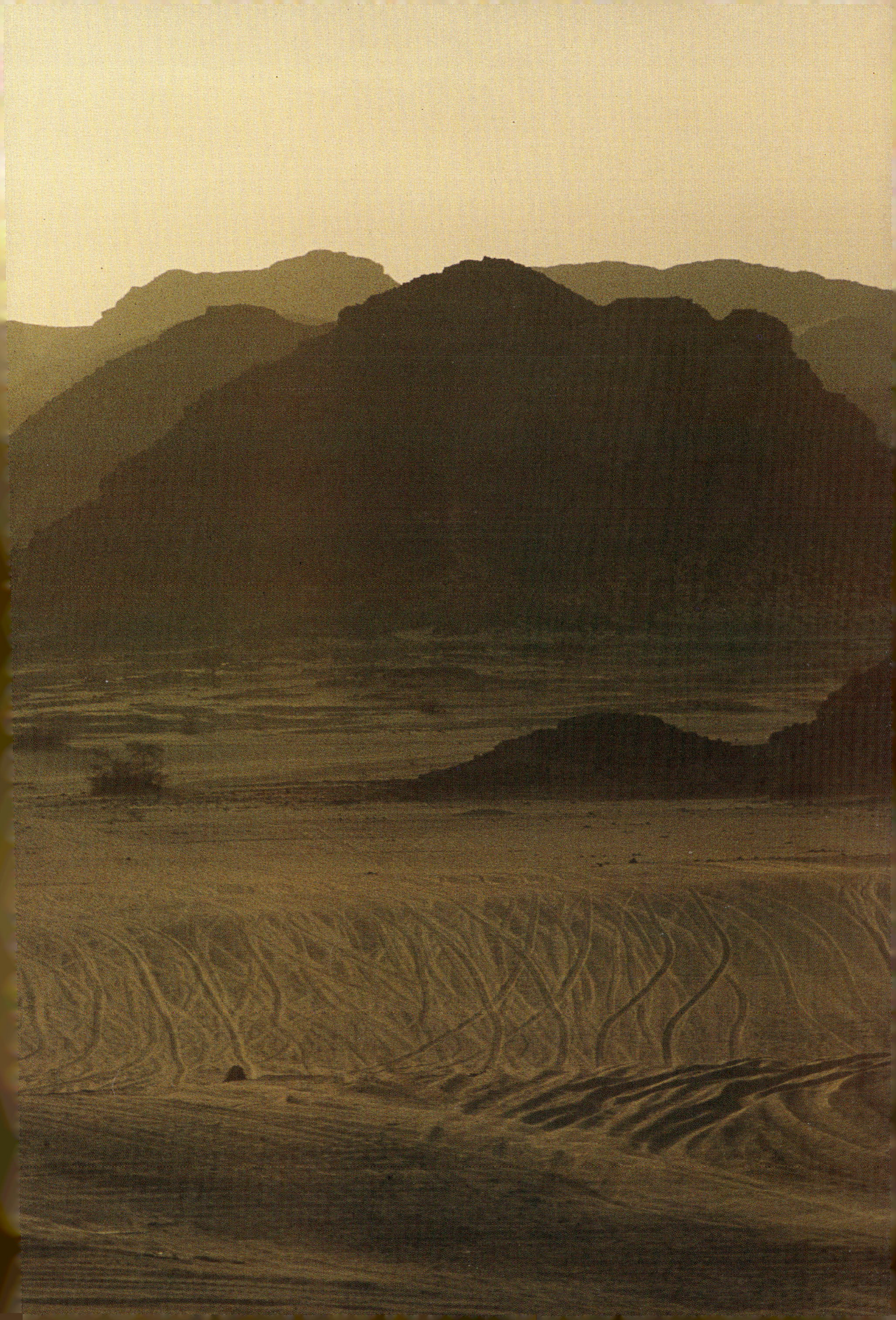

△ The little garage at Djanet where we repaired our trackrods and the Norwegians welded up their chassis

◁ Djanet

◁◁ The pick-up, steering on just one wheel, on soft sand before Djanet

Our camp outside Djanet
▽

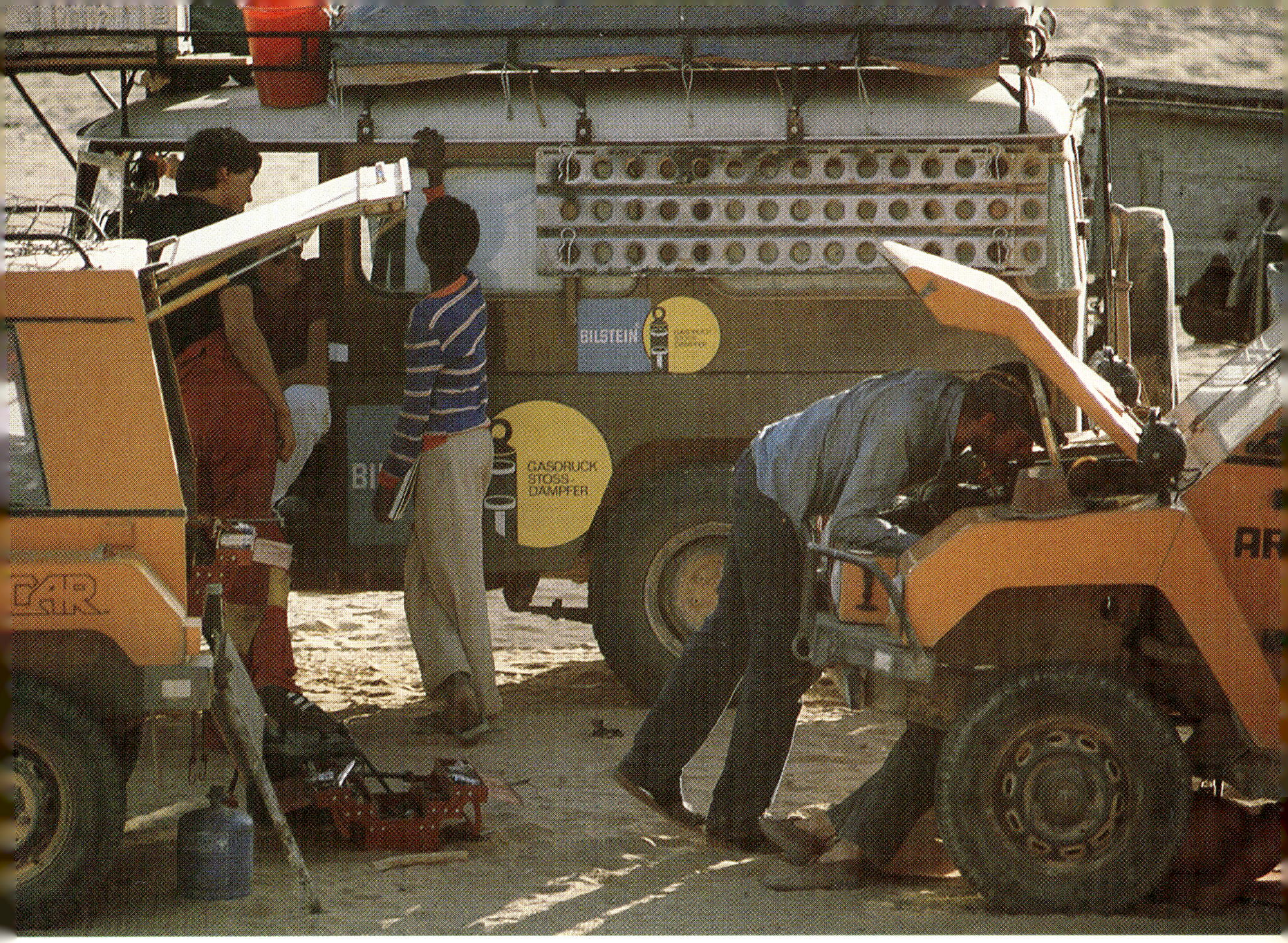
BILSTEIN
GASDRUCK
STOSS-
DÄMPFER

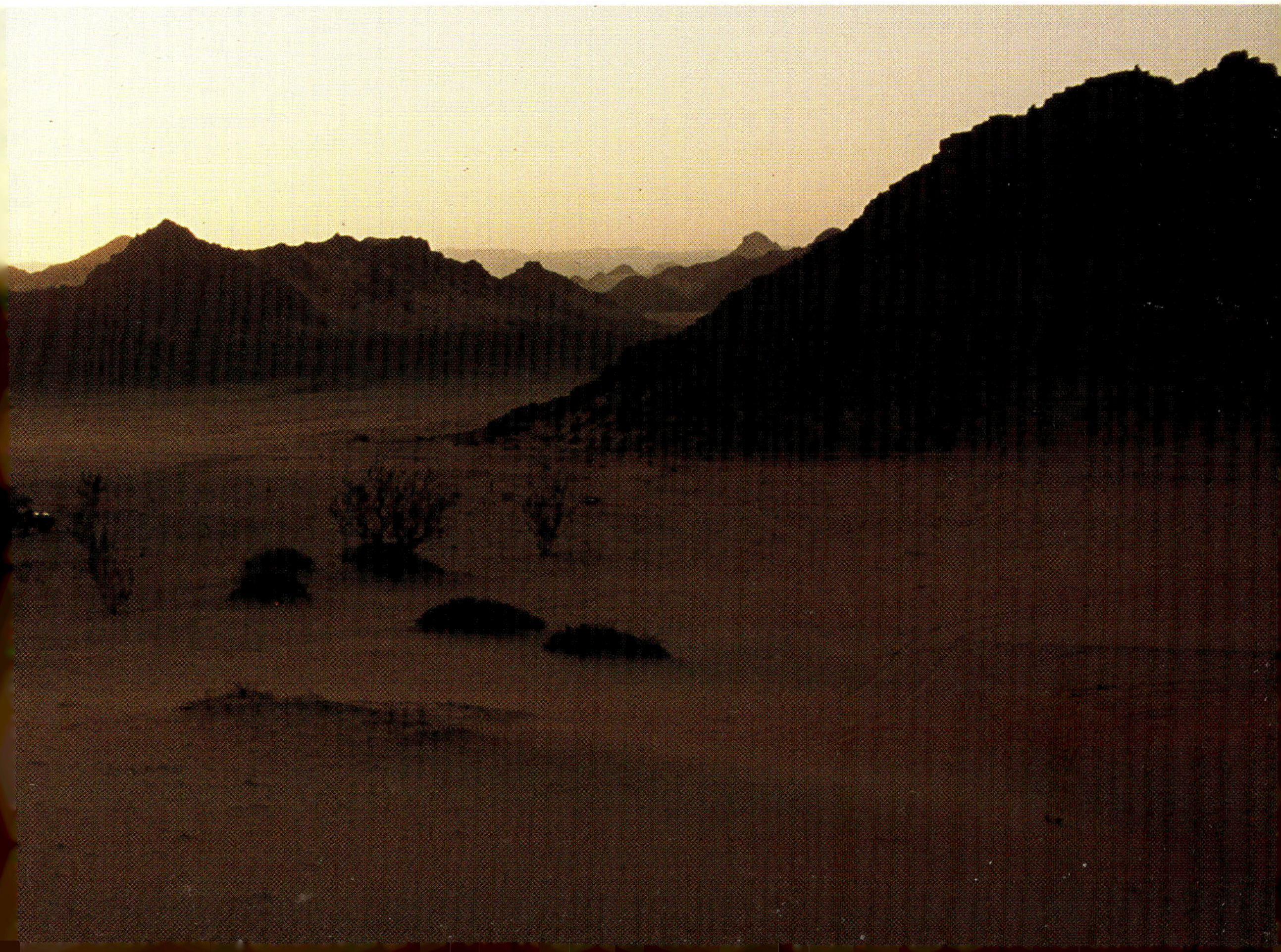

Tassili, always beautiful and always changing

The first gearbox re–build at Fort Gardel

Heavy trucks smash their way at speed across the Sahara

Travelling with the Toyota emphasised for us the value of careful, practical design.

The pick–up is doing about 28 miles per gallon or 10 litres per hundred kilometres in the desert. It is carrying fuel for 1,500 kilometres and is loaded with a hundred litres of water, two batteries, two spare tyres, a wide range of mechanical spare parts, food, cooking facilities and personal effects and sleeping accommodation for two people.

The Land Cruiser is carrying about the same, yet it looks a mess. It is cluttered with extra jerry cans, sand ladders and other bits and pieces. Loaded it has about twice the frontal area of the pick–up and uses about double the amount of fuel. Its limited ground clearance severely restricts its off-road capabilities.

Our next objective is the massive Ahaggar Mountain range. The climb into the Ahaggar is on a route designated four-wheel drive only. It is 1 in 3 and steeper with a loose and rough surface.

Tony Hughes waits for the gearbox to be repaired ▽

The climb into the Ahagar Mountains on a ▷ four–wheel drive only route

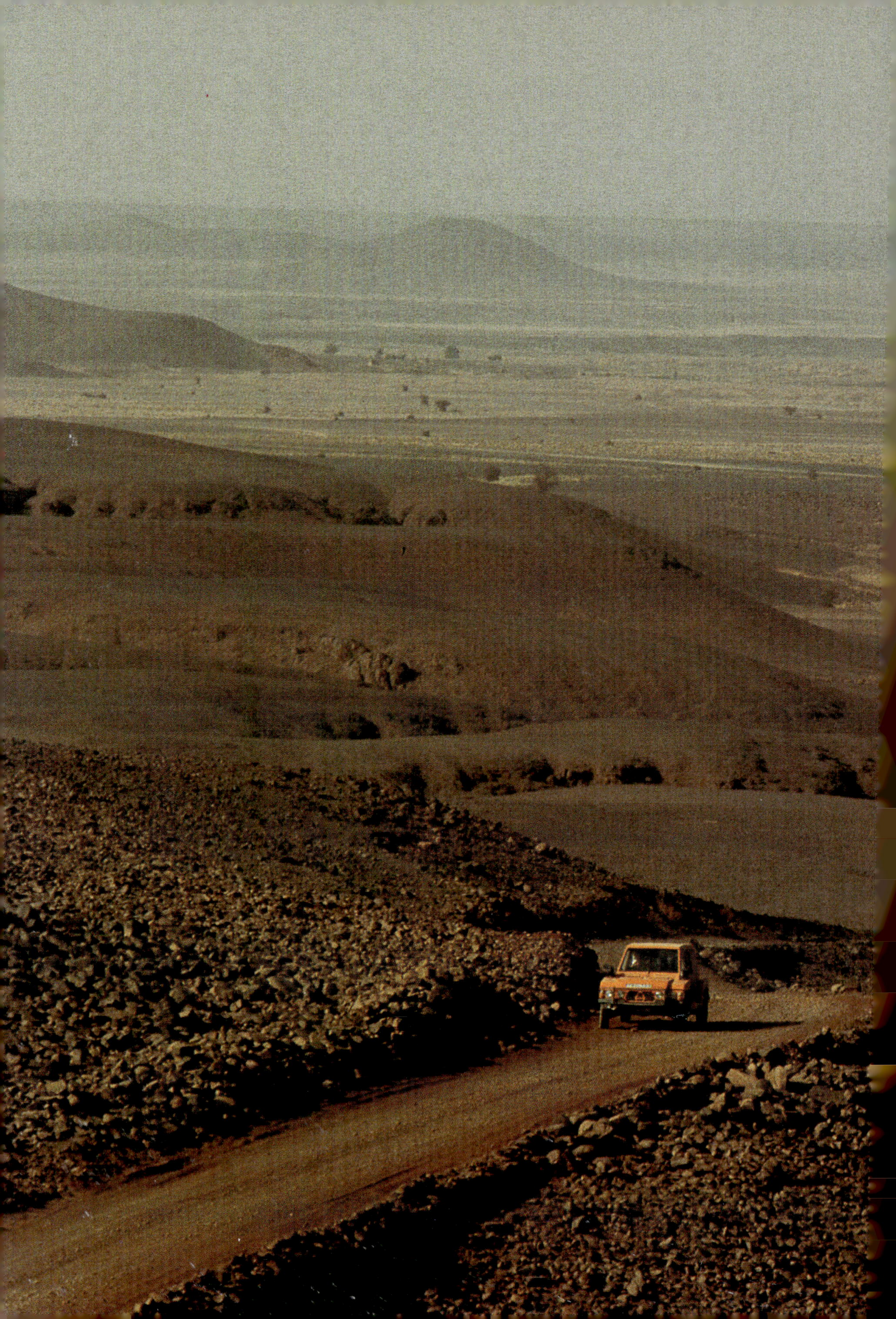

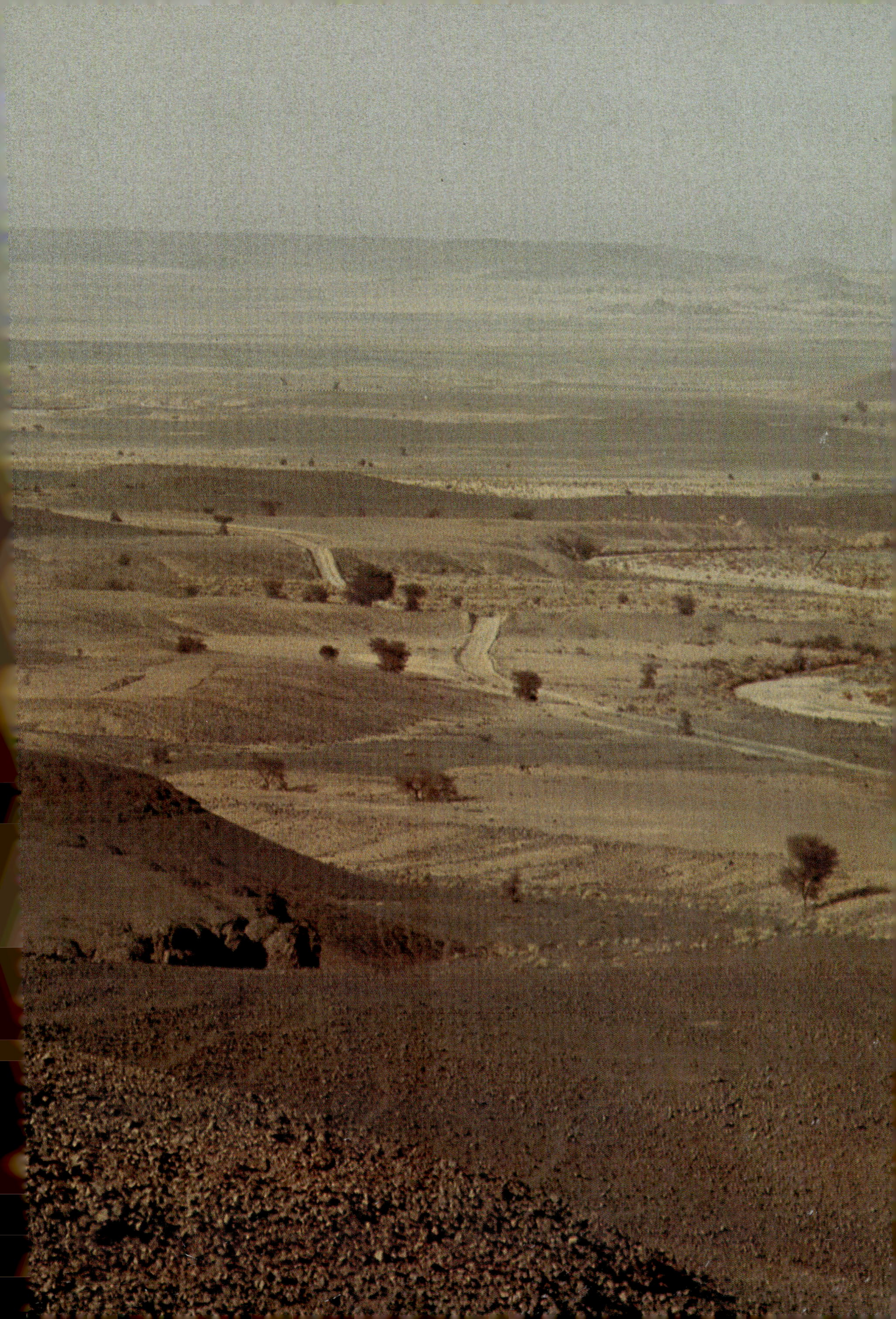

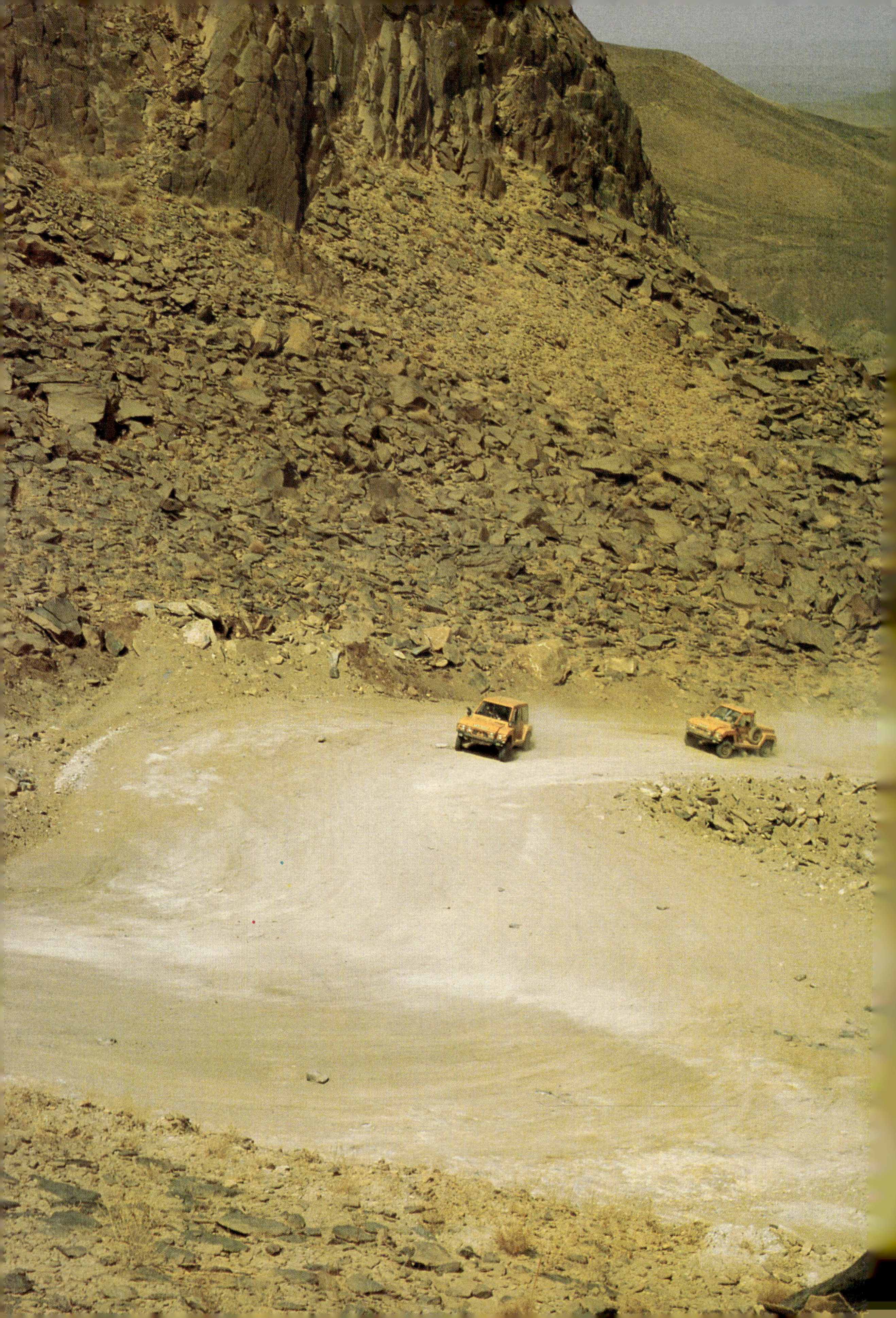

The six–wheeler is stretched to its limit. With, at this development stage, just 65 horsepower and an unsuitable gearbox, it has to be unloaded on some sections.

The six–wheeler's nose lifts as a reaction to the torque being transmitted by the front driving wheels, but it does not lose adhesion.

The altitude starts to affect engine performance. Without the luxury of a low ratio gearbox it is necessary, on the final climb, to slalom across the road to reduce the steepness of the ascent.

At the top we encountered the inevitable 2CV Citroen, overloaded and only just able to carry its own driver.

Ever since people began to understand the principal of the lever, almost all engineering has been derived rather than original.

In the case of the AFRICARS, the 2CV or Deux Chevaux, has been the principal source of inspiration. This mid 1930s design is astonishing. Most modern vehicle design does not even address the problems solved in the 2CV. It has rightly been called: 'The only car of the future that we have'.

Respectfully we watched it pass and proceed calmly and steadily on its way to Djanet – undaunted by the four–wheel drive only route.

We pressed on to the heart of the Ahaggar intent on seeing the sunset from the top of the 10,000 foot peak, Asekram.

The inevitable Citroen 2CV
▽

The six and seven volcanic plugs that make up the structure of the Ahagar mountains are eerie and ancient
▽

We are out of Algeria and back in no–man's land. Bob has gone home with the Toyota and Thomas Marx is driving the six–wheeler.

Now we are amongst the traffic of the main trans–Saharan route with the AFRICARS leading the pack. They demonstrate that they can more than hold their own against traditional four–wheel drives.

Our driving on soft sand has become more confident. The ability of the cars to ride out the unexpected bumps and holes means that we can set just the right speed.

Watching the antics of the flying Peugeots makes us a little smug. They have to use unreasonable speed to get through and even then they don't always succeed. One had to be jacked up and dug out eight times since Tamanrasset and it wasn't taking kindly to the violent undulations of the piste.

We lost another two weeks in Tamanrasset because the border to Niger was closed. Now it is open we are making the best time possible. Another forty miles of untended no man's land and we will reach the frontier.

The worst of the desert is behind us – possibly the hardest route that we could find anywhere in the world. The cars are complete and running well.

The desert in Niger has changed character completely. It is flat and the cars are rolling beautifully, the suspension is working really well on the desert piste. We are able to travel 60 kilometres an hour without any trouble as we head towards the capital of Niger, Niamey and the river Niger.

We have entered the Sahel – the arid sub-Saharan region in which life is tenuously maintained by water from wells hundreds of feet deep.

For me the desert was no different from the first time I crossed it twenty years ago. It's very nice to be out and yet from my own experience when you drive north through Africa the desert itself becomes a release from the pressure of population, of mosquitoes, flies, the oppressiveness of the jungle where you have a vast landscape but you can't actually see for more than a few metres to the side and a few hundred metres ahead.

The desert is a release, the nights are cool and it's not humid. Right now getting towards equatorial Africa seems a release from the pressures of the desert. It gets you in every way.

THOMAS: 'It's a little frightening out there in the desert and it is a good feeling coming to the Sahel. Seeing the trees, the people and the animals, the cows. They are alive there is all this life around you.

'It is a good feeling knowing you are somewhere where you can stay and relax. You cannot do this in the desert, all the time you are aware that you have to keep moving to survive.'

CHARLES: 'I can remember the enormous difference in the speeds at which we've driven. Coming into Geneva we clocked 134 kilometres an hour in the six–wheeler. To think of doing that now seems complete insanity.

'When we came into the desert the conditions got worse and worse and we started to go slower and slower. There were days when we were in the rough stuff and we were driving between five and 15 kilometres an hour all day.

'It was actually quite nice as it gave you the time to look out of the window and see what was going on. You didn't miss very much.

'Then we got to Niger and it was all there – tarmac stretching as far as you could see and you could just accelerate up to 130 kilometres an hour. But as soon as you got up to about 80 you just began to feel unsafe. There were cattle, sheep and goats wandering along the sides of the road.

'You realise how fast you drive in Europe and how you get accustomed to fast driving and yet it really isn't safe. No–one can anticipate what the animals on the side of the road are going to do at that speed. I think it's going to take quite a while until we're actually happy driving at those sort of speeds again. Maybe I'll never be happy again driving fast. I don't know.'

At the frontier post we heard a rumour that now the Nigerian border is closed after a coup, so we have decided to abandon our original route south to Kano and instead, to head to Niamey. There we can replace the pick–up's windscreen and see what, if any, routes are open to us.

A deep Sahel well in Northern Niger ▷

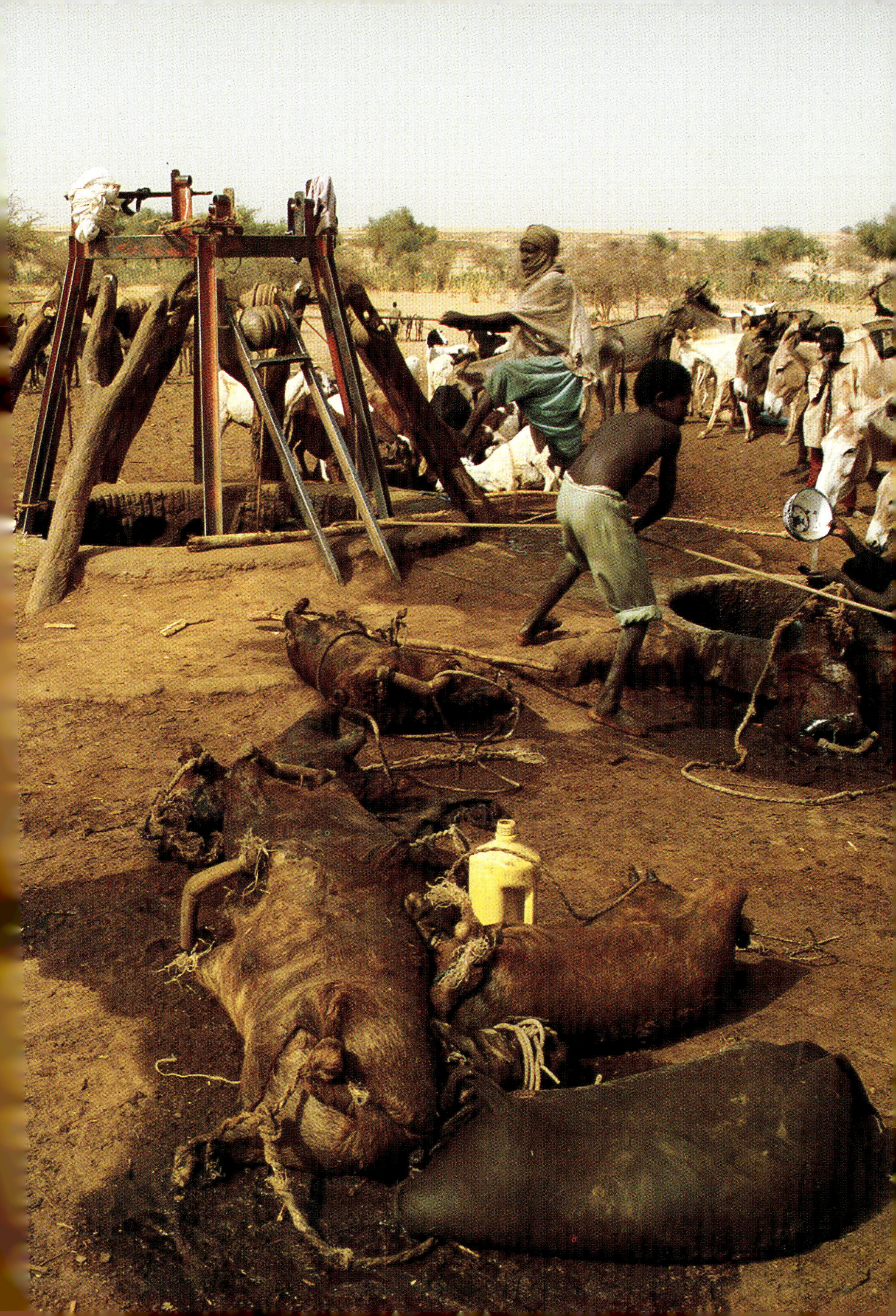

The tarmac is astonishing. The road is the uranium highway, a strategic road built to get uranium from north eastern Niger, the first stage of its journey to France.

On arrival in Niamey we found that the Nigerian border was indeed closed and we had no option but to book into a hotel and make a new plan.

There is something obscene about sitting beside a swimming pool in a luxury hotel right in the middle of the region known as the Sahel – the band across Africa which probably at this moment has the highest rates of malnutrition, the greatest poverty and the greatest number of people actually starving to death, depending on how the rains come and go.

In a curious way it is a luxury area. As you come out of the desert there's the luxury of seeing deep wells and water, sleek looking cattle, very well cared for villages, even tarmac roads. It somehow doesn't seem like an

Neat and ecologically efficient village in ▷ Niger

The Uranium Highway
▽

impoverished area, it's almost like a fantasy land, a fairyland. And yet the hard economic facts and the problems of maintaining life are very real.

CHARLES: 'I can remember the three images that pass through your brain while lying in the car feeling pretty miserable and very, very hot.

'One was diving into a swimming pool, the other was having an ice cold drink and the last was being able to recline in an air-conditioned room. Of course arriving in Niamey where we have the lot is like coming to paradise, really nice.'

CAROLYN: 'I had been in deserts in California and the Western United States and I liked them. When I came to the Sahara, the desert was the desert as I know it but so much more. It was incredibly overwhelming – to be so tiny, just a human and not to make any difference to the environment at all. It felt very good to be so insignificant. I liked it.'

LIZA: 'The desert was empty and there

was so much space but I never felt completely alone. I suppose on the routes we were taking there was quite a lot of traffic. You may see only one car a day but somehow that was very reassuring.'

CHARLES: 'I can remember walking away from the camp and just going behind a sand dune so that the camp was out of sight and you felt you were in the middle of an enormous wilderness. It was quite frightening actually, for a minute or two.

'You eventually became used to it and after a few days you began to resent people coming along and intruding. It was rather nice to be on your own in such a wide open space.

'I remember another time when we were driving along a very wide open valley and the two tracks split, as they quite often did when a new track has been formed around an obstacle – occasionally they never join up again.

'We ended up after half an hour with two of the cars about three or four miles apart –

△
The River Niger

I think. We stopped and stood on top of our car and scanned the horizon with binoculars to try and find the other.

'It was only because of the bright air–sea rescue orange that we managed to pick out the AFRICAR. It was tiny, quite unbelievable to think it was so far away. The sheer scale of being human in a place like that makes you feel tiny.'

Tony Hughes was sent on ahead, charged with flying to each capital city along our projected route to check and confirm our permission to travel and to film.

But Africa does have a way with plans. He was robbed at Lagos airport and lost all of his documents and a critical portion of our precious funds.

The next thing we heard he was home in Nairobi and we were on our own.

For the moment we feel quite encouraged and ready to set off again. Once more we expect the primary problem will be getting across borders. For the moment the Nigerian border is closed but it should be open in a few days.

Any apprehension we have is whether we can get in and out of countries with our cars and our equipment. We want to be able to travel peacefully while just fighting the road, physical and climatic conditions – but not the political ones.

In Niger we are about half way through the African part of our journey. Reduced to just five drivers and with no desire to back-track through the desert, we are committed to continue, travelling hopefully.

But with a war in Chad and Nigeria closed, there seems to be no open route to East Africa or even to the Equator.

6
SAHEL

The first of two trees under which we spent almost a week in Niger waiting for the Nigerian border to open

With Carolyn and Marcelle trying to decide on a route

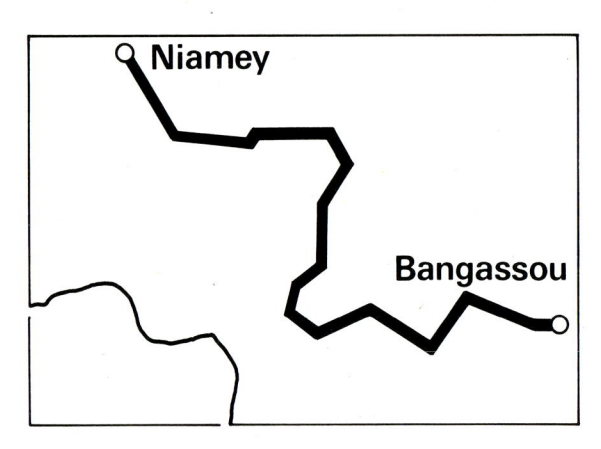

Africa has few paved roads, but in our experience most of them are in Niger. Over a period of seven weeks we drove more than 6,000 kilometres backwards and forwards over the magnificent tarmac roads of this sub–Saharan country. But roads are not a lot of use if they don't lead anywhere.

In Niger we found ourselves in a situation which seemed so unlikely in the 1980's and yet, in reality, is a disease of the 20th century. The passport is a relatively modern invention and it has a lot to answer for. Before the passport the world was a far easier place to travel around. Admission to a country depended to an extent on whether people liked you or wanted you. Today, without a passport it is virtually impossible to travel, and in our recent experience, even with a passport it can be impossible to travel.

The reason that the Nigerian border was closed to us, and to everyone else, was that a coup had taken place. A new government was installed with the intention of wiping out the unbelievable corruption that had developed in Nigeria during its brief period as a relatively rich oil producing country. One of the problems was that vast quantities of money, running into billions of dollars, was frequently being smuggled out of the country into Europe across borders such as the one we wanted to cross. This money had all been illegally obtained and often was part of deals with Western governments or Western financial institutions, which most of us would presume, should know better.

To stop this, the new government decided to change the currency of the country. This meant that by a specified date people could take their old money, prove its legitimacy, change it for new currency, and then carry on life as normal. Those who had amassed fortunes of illegal currency would not be able to exchange their hoard. So the borders were closed to prevent any currency transactions going in and out of the country. This extended beyond land borders, even airports were closed.

There was a further problem. With the declining price of oil, Nigeria's fortunes were on the wane. For a period of time imports had been increasing at a rapid and unreasonable rate. By closing the borders imports were simply stopped while exports, consisting primarily of oil, could continue. The problems of Nigeria were certainly partly self–inflicted, but they were aided and abetted by people in the City of London, Wall Street, in Paris and Switzerland with an eye on the main chance of making money and effectively ripping off yet another Third World country.

Typically, advantage was taken of the relative prosperity of Nigeria to increase the level of imported manufactured goods, while at the same time every effort was made to reduce the price of oil. Another curious and self-defeating equation.

We, and many others, including the cattle ranchers of Niger who depended on exporting cattle to Nigeria for their livelihood, were caught in the aftermath.

Our original route via Kano to the Cameroon Republic and Central African Republic before travelling on to Nairobi has had to be abandoned.

The possible alternative seems to be to drive as far east as possible in Niger and try to creep around the south of lake Chad through a kind of no–man's land, without being detected.

We could also take a route north of the lake across the massive dunes of the Great Erg of Bilma to N'Djamena, capital of Chad, and then south to the Central African Republic.

There is one snag. This route is right through the middle of a war between Chadean factions backed on one side by Libya and on the other side by France.

We can't afford to stay on in Niger. We have already stayed more than a month

◁ To life and friends under a tree

we get to the corner of Cameroon?'

CHARLES: 'I don't know. If Nigeria is closed, Africa is closed. I can't imagine us sneaking past Lake Chad without someone spotting us. There's always going to be somebody peering through a pair of binoculars wondering what we are doing.'

There could be an advantage to that if we had a problem. There is a sort of theory about deporting people that they send you the shortest way out of the country. It's possible that we could be deported to the Cameroon.

LIZA: 'What happens if the Cameroon border is closed?'

I think that is strictly rumour at the moment, I don't see why that should be the case.

THOMAS: 'I heard from the Swiss in the Grand Hotel that they waited about three weeks to get into the Cameroon from Nigeria, and the reason they couldn't get a visa was the closed border. They had to spend an extra two weeks in Nigeria and only managed to get out with a lot of trouble.'

Anyway, we'll have to give it a try because there are barely any alternatives left. We could possibly go through Chad, right around the lake, if we can get permission.

THOMAS: 'There are big dunes over in Chad and I don't know if the AFRICARS can manage them. There is an unmarked piste along the dunes, but you only have the lake for orientation. At times you travel a long way from the lake which makes it extremely difficult.'

It's the sort of place you would only go with a Unimog and even then you could run into trouble. Fortunately the political situation in Chad appears to be calm. I think the French and Libyans have reached a kind of truce. The Libyans said that they would go if the French leave and the French said they were ready to go in twenty-four hours, that was the last I had heard about it. The border is open, but again we have to get permission from the Niger side to actually travel there because it's through a military zone.

The other alternatives are almost unthinkable. One of them is to sit and wait for the border to open. That could take a week, six months or five years. We could possibly go

and covered an additional 3,000 kilometres exploring possible methods of escape. Our budget shows that we do not have enough funds left to reach Nairobi.

We have had to leave our hotel in the capital of Niger and now, camped beside the road to Chad, we must make a decision.

CAROLYN: 'What about going up to Lake Chad, then instead of trying to get down into Nigeria, just keep sneaking along until

131

down through Benin and take a boat to Douala in the Cameroon.

CAROLYN: 'One thing about the rumours we have heard is that hardly any of them have been accurate. You really have to go in there and find out.'

Yes, you have to travel, hopefully, and try to get information on the spot, then follow your instincts. Put a bit more tea in the pot and then we should get going. Once again it's a nice day.

We eventually decided that it was too difficult and dangerous to go north of the lake, but we had to at least make an attempt to get around the bottom corner or even across the water, by raft if necessary.

As we approached Diffa, the last frontier post to Nigeria, we met Marcel, travelling on his own in an old 404 Peugeot. He explained that he had already tried to get through at Diffa but, after being given permission to leave Niger, he was stopped by the Nigerians.

Marcel was convinced that the border would be open by the 26th of May, in just three days. He warned us that we couldn't get close to Lake Chad without being detected and if it rained the tracks would become completely impassable.

The Nigerians hadn't given him any trouble but they had told him that the information about the border he'd been given in Paris was for Paris and not for Nigeria.

Marcel lives in the south of France and runs an unusual vehicle export business.

Our local well

He makes eight trips a year across the Sahara, delivering huge multi–wheel drive Mercedes trucks.

Each trip he packs the trailer with three second–hand Peugeots. Two are sold with the truck and one kept as his own transport until he is ready to fly back to France, then he sells it.

Eight Saharan crossings a year suggested that he might know something that we didn't. So we decided to go along with him and find a tree to wait under for three or four days.

We back–tracked towards a small frontier post, still some 1,200 kilometres from Niamey.

Before approaching the border zone we filled every available container with water – just in case. Then Carolyn went to look for a tree.

It turned out to be a magical tree, surrounded by leafless baobabs, the oldest and perhaps the ugliest trees on earth.

Filling up with fuel before Diffa and our final departure from Niger
▽

For the first time since leaving England we had neighbours and became part of a community. Our presence with the three orange cars just seemed to be accepted. Hospitality was extended, presents exchanged and for a few days life became almost normal.

None of that 'frightening isolation' as Thomas had said of the Sahara and no compulsion to get on the road and keep moving every day.

We were camped on the edge of a small farm where there appeared to be nothing except sand for soil. All the water for drinking and irrigation had to be drawn by hand from a well 70 feet deep.

The further we had travelled the clearer the required specifications for the two missing AFRICAR components, the engine and gearbox, had become.

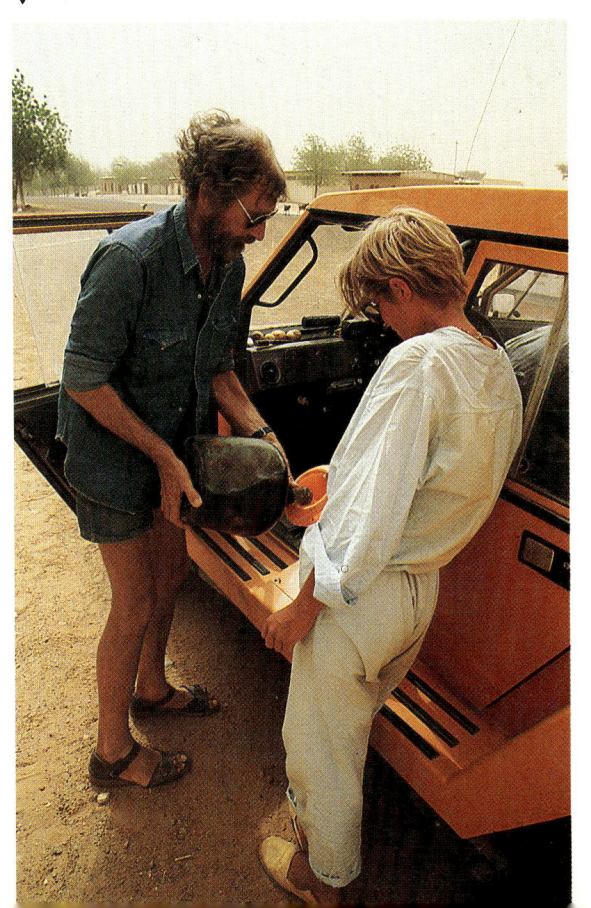

Thomas and I had hours to spend bouncing ideas around and making preliminary sketches on a small drawing board, brought from England for such an opportunity.

With the border closed to Nigeria, the nearest oil producing country, fuel was getting scarce in Niger. Our tanks were empty after the long run from Niamey.

But the bush telegraph worked well and 400 litres arrived at two o'clock in the morning, carried in 40-litre containers on the heads of a seemingly endless line of porters. Prices were competitive and the source unknown.

It looks like the information from the border this morning means that we are stuck on this route. The Nigerians suggest that the border may not be open for another month. Even if the border opened now, we are running very close to the possibility of hitting the rains in Central Africa and Zaire. If that happens we could be stuck there for two or three months as the roads become impassable.

◁ **Out of Niger into the wasteland of Chad**

A little corner of Lake Chad
▽

I think it has become impossible for us to carry on with our original route. For the moment we really have two choices. One, we head straight back across the Sahara to Algiers and ship home that way and the other is to travel down the coast to Togo or Benin and from there go back to England.

MARCEL: *'If you are still here next month may I meet you? I have a truck ready and if the frontier is open I will come back. Good luck and Bon Courage!'*

Watching Marcel go and knowing we had to leave the tree was quite a wrench. With little prospect of getting on the road again, at least in the right direction, it seemed perverse to leave such a friendly, comfortable and secure place.

The good thing about the past month stuck in Niger is that the people here, the authorities, in fact everyone, have been so helpful and

△ **Probably lost**

Begging water somewhere in Central Chad ▷

◁ **Preparing to drink the green water of Lake Chad**

pleasant to be amongst. It's been a very good African experience for all of us.

It's a curious thing that in 1984 it is no longer possible to travel overland between major cities, not only in Africa but all over the world.

Today, to make the same trip that Marco Polo made from Venice to China, would be completely impossible. Right now you cannot get from northern to southern Africa. The Sudan is blocked as a result of a civil war and you cannot get visas. The Nigerian border is closed which means you can't travel from West into Central Africa.

The same thing is happening in Central America. It really is curious that in the age of communication, computers and aeroplanes, when in theory the world has become a much smaller place, it has also become a bigger place in that you cannot travel overland.

We had discovered that even if we wanted to go north to Chad, we would have to go back to Niamey, an additional 2,400

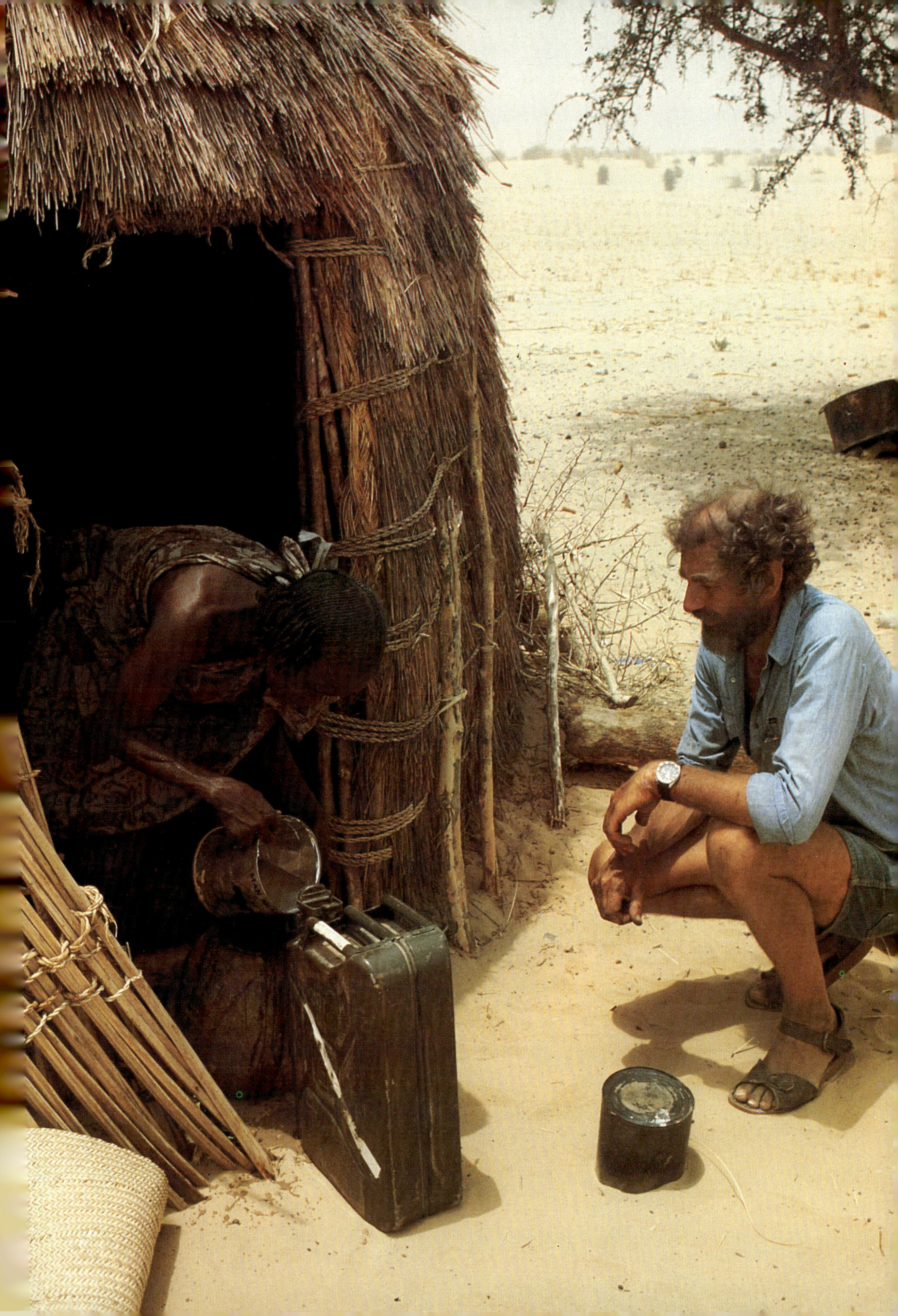

kilometres, to get permission to pass through a military restricted area on route to the border.

On the drive back I realised that we actually had no affordable alternatives. Our return tickets were from Nairobi and non-exchangeable and we did not have the finance to get home any other way, with or without the cars. Our only real option was to go forward – overland.

Optimistically, I still presumed that if we succeeded in completing the trip and doing what we had predicted we could do with the AFRICARS, the financial pressures might ease.

So I had resolved to get the required permit and head for Chad, war or no war, big dunes or the rainy season. This had to be a succeed or bust attempt to get through.

The AFRICARS may only be prototypes but they have been designed to cope with all African conditions. And cope they better had!

The permit to leave Niger in the direction of Chad was obtained in three days, but only on the condition that we used it entirely at our own risk and responsibility.

Now we were navigating by compass bearings and following the occasional random tracks. We had no visas and had passed no border posts, which was probably fortunate as Charles had lost his passport.

The main thing was we were moving and at speed. Our only concerns were getting lost with no one in the world knowing where we were and the probability that the little temporary gearboxes would not hold up under the continual drag of soft sand.

The Sahara was just a practice run compared with this. These dunes are huge, they don't even look like dunes and the sand has the texture of talcum powder.

After crossing the border into Chad the signs aren't exactly encouraging. We are starting to get punctures from large thorns and we are also driving through what has until recently been a war zone in the dispute between the northern and southern factions in Chad. Right now it all seems very calm.

Another gearbox re–build somewhere to the North East of Lake Chad
▽

The route we are taking is not the normal way which goes far north of the lake. Instead we are crossing what used to be the northern edge of the lake, it is not easy. The water has receded and the part which was the bed of the lake is quite hard. It is when you get to the sand dunes that run like spits down into the lake that it is difficult. They are very soft but the car's are managing it.

We have covered about 1,600 kilometres in the last four days and we hope to get that up to about 2,000 by this evening. That puts us more than a third of the way to Uganda.

This route is a little uncertain and we just have to proceed as an act of faith. All we know is that the route is going east and that's the direction in which we need to go.

We are getting a little tired of rebuilding these small Citroen gearboxes. They have done very well because they are pulling twice as much torque, twice as much power, than their original design specifications. In fact, the one in the wagon has come all the way from England with only a single change of the small bearing.

The gearboxes in the six–wheeler and the pick–up have now been rebuilt five times and this is becoming increasingly difficult to do in the desert.

The gearbox has absolutely nothing to do with the AFRICAR. It's just a means that

△
The six–wheeler showing its pace, travelling flat and fast regardless of terrain

Riding the top of the massive sand dunes ▷ north of Lake Chad

△
Facing virgin countryside with no defined route. Which way do we go?

we use at the moment, while we're waiting for our own box to be ready. Now, 100 kilometres short of N'Djamena, we have no option but to settle down once more under a tree in the desert and get ourselves mobile again.

Nothing, not even broken gearboxes, could detract from the excitement and pleasure of our journey through Chad. We all loved it.

Driving free through virgin country is an unusual experience these days. Breasting a dune and seeing nothing in front except wave upon wave of sand was slightly frightening and yet so unique as to be unforgettable.

There is no way we could have tested for these conditions in England. The circumscribed fields and tracks of a farm near London are pitiful in relation to the scale and scope of Africa.

But a bump or a ditch is the same in Europe or Africa. It was good to feel the security of the development of our suspension from the nose–down landings of the first prototype to the flat and controlled ride in Chad of the later cars.

The landscape was extraordinary. Sometimes the valleys between the dune peaks

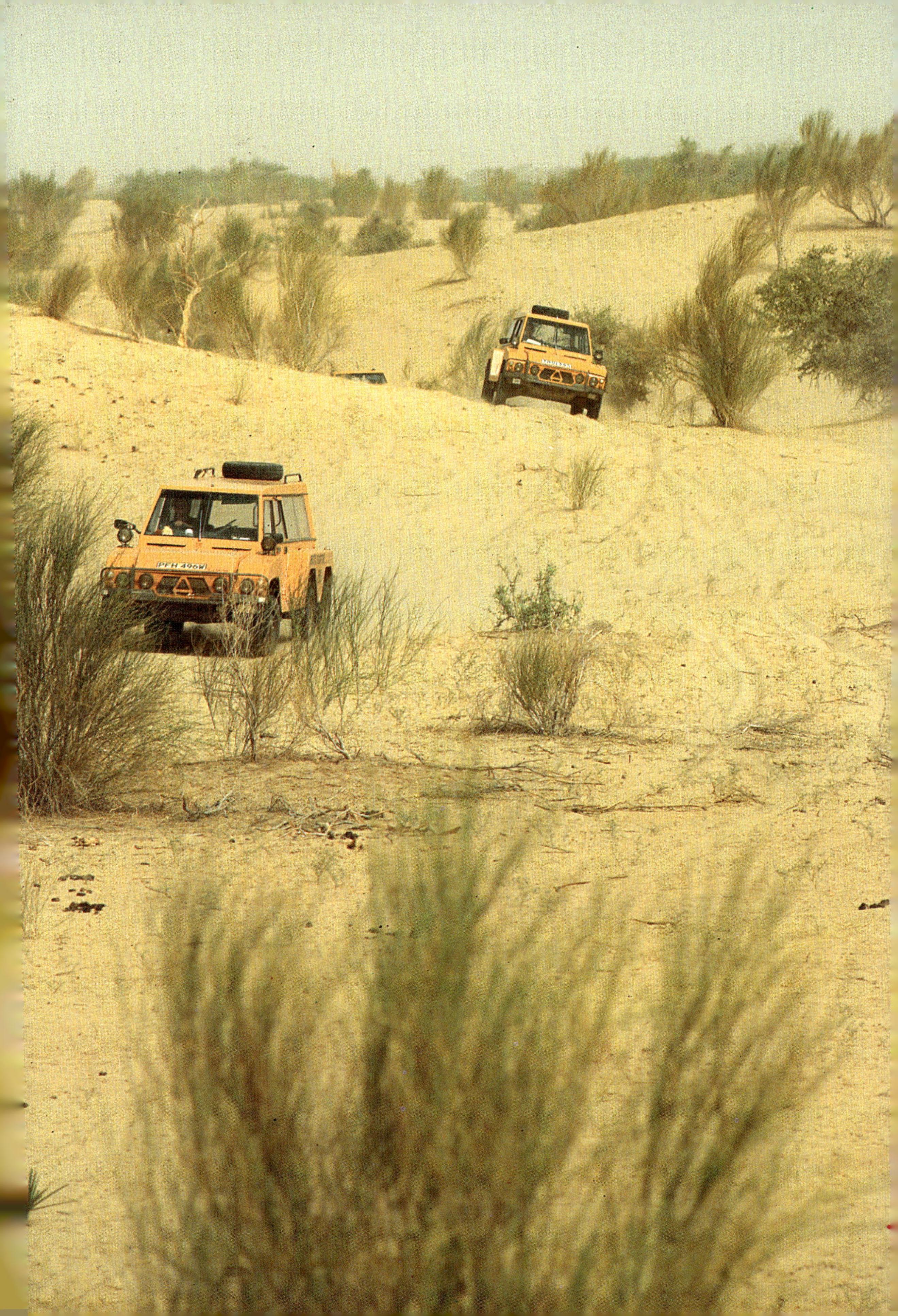
PFH 496W

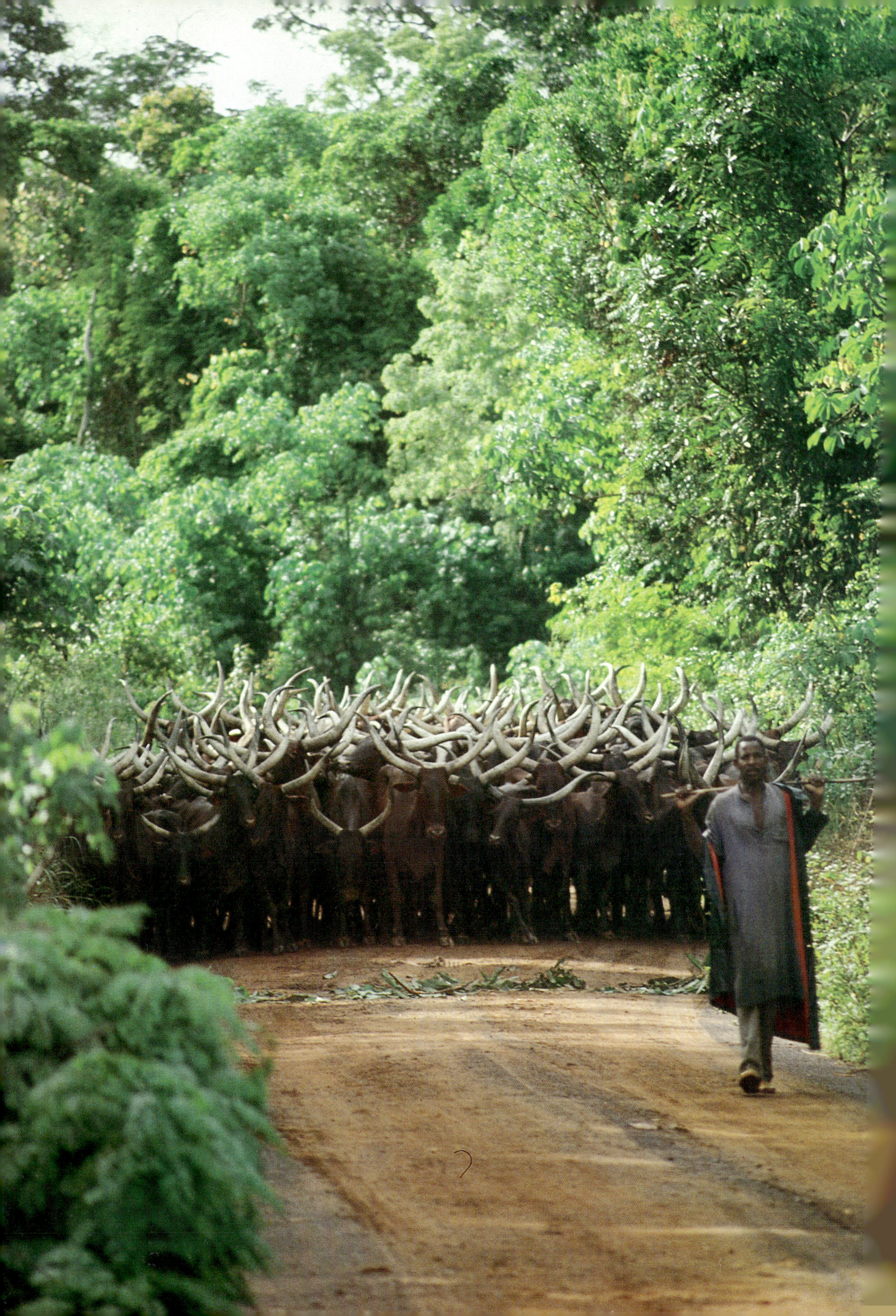

In Southern Chad we sought the security of the villages because of continuing warfare ▽

The perfect AFRICAR road in Southern Chad ▷

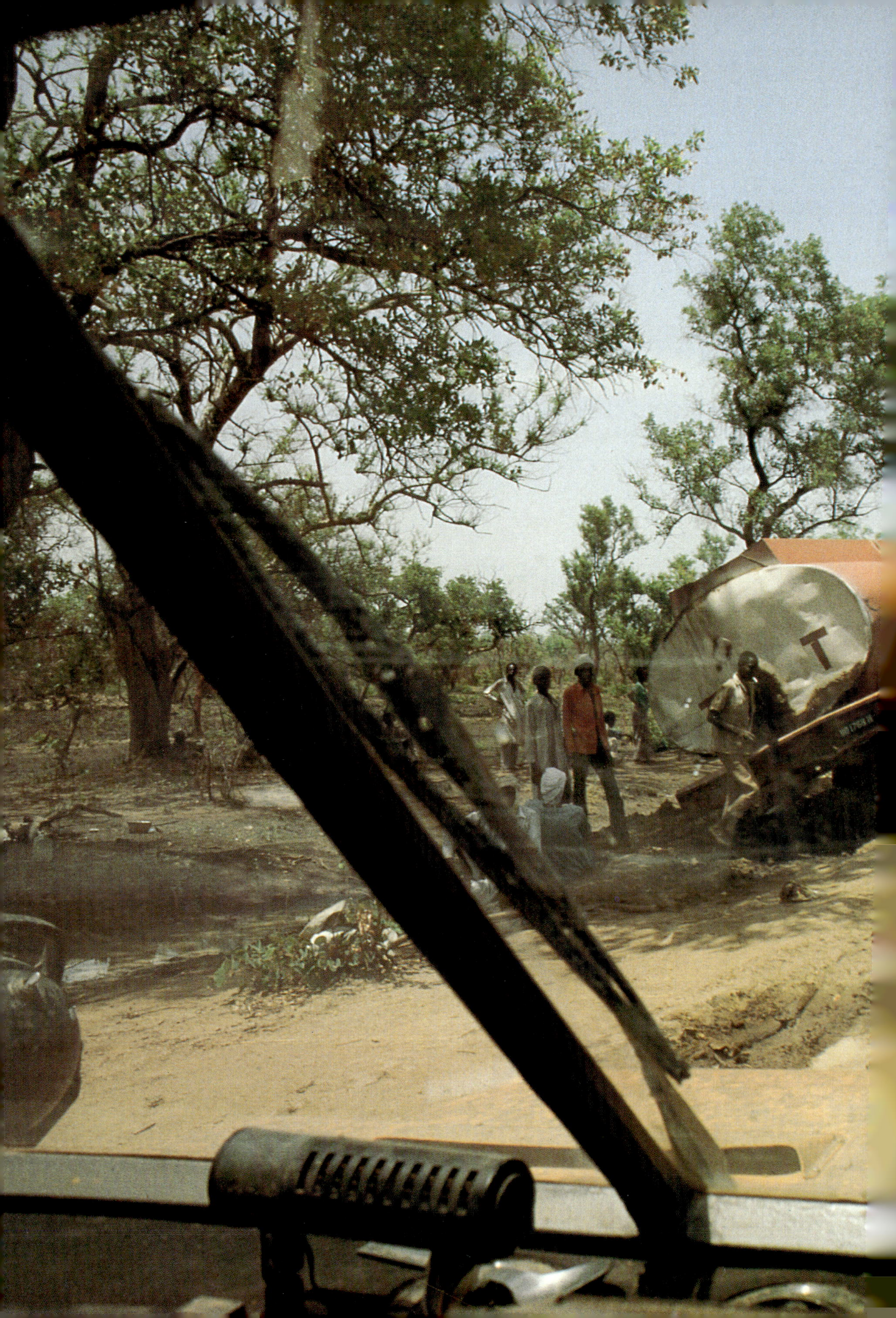
T

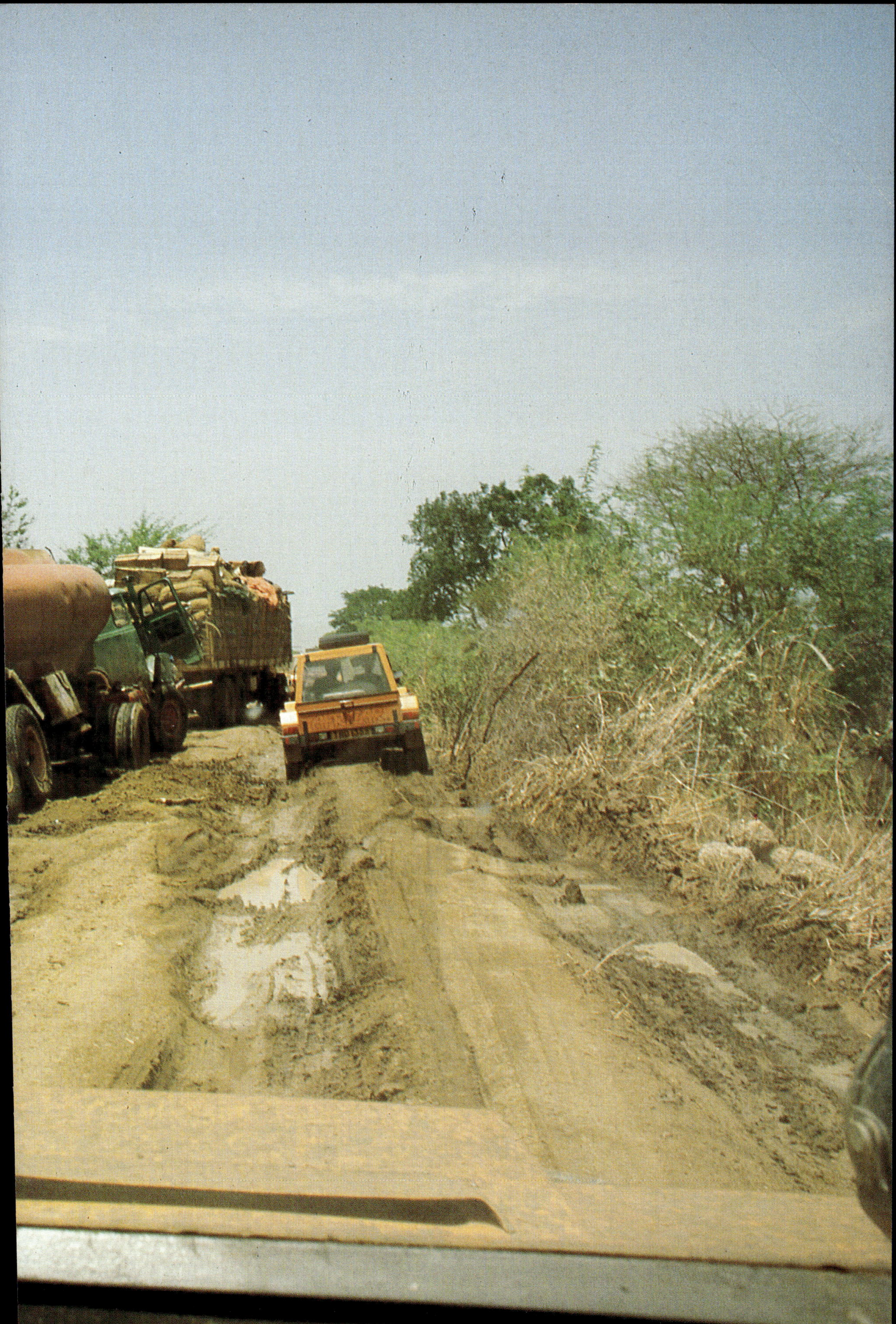

One of the many small police checkpoints on the road from West Africa to East Africa

Temporarily under arrest in Sarh in Southern Chad

resembled classic English parkland with carefully positioned trees and manicured sand passing for mown and rolled grass.

The soft peacefulness of the landscape of war–torn Chad only needed water and greenery to make it Eden.

There were no problems for us in shell–battered N'Djamena. We were welcomed, with a few raised eyebrows, as tourists, given visas in half an hour and had no difficulty with the missing passport.

As we underwent the rapid transition from North Africa to Africa south of the Sahara, we seemed to enter an easier and more natural world, less bureaucratic and perhaps less dogmatic.

In N'Djamena we obtained updated visas for Zaire in half a day without fuss or cost other than the standard visa fee.

We are heading south on one of the two main supply roads for land–locked Chad. There has been a little rain and the heavy truck traffic is starting to break up the road. If the rain continues this road will be blocked for weeks.

In southern Chad skirmishes left over from the war are still quite frequent, so we seek the security of villages at night.

We've crossed the border into the Central African Republic and are in the beginnings of equatorial jungle. Mosquito nets are now essential for malaria is rife.

The roads throughout Central Africa are of laterite, a hard, red iron ore based clay that makes a good durable surface, if maintained. Once it's allowed to deteriorate the rain will wash away patches and then, with no hard foundations, the trouble starts.

At Sibut we stopped at an American Baptist Mission. We had heard they had the only welding plant for hundreds of kilometres and we wanted to check all critical components before the rains started in

Refuelling at Sarh before crossing to the Central African Republic
▽

The tiny band of green open country was refreshing after the desert and before the jungle closed in the Central African Republic ▷

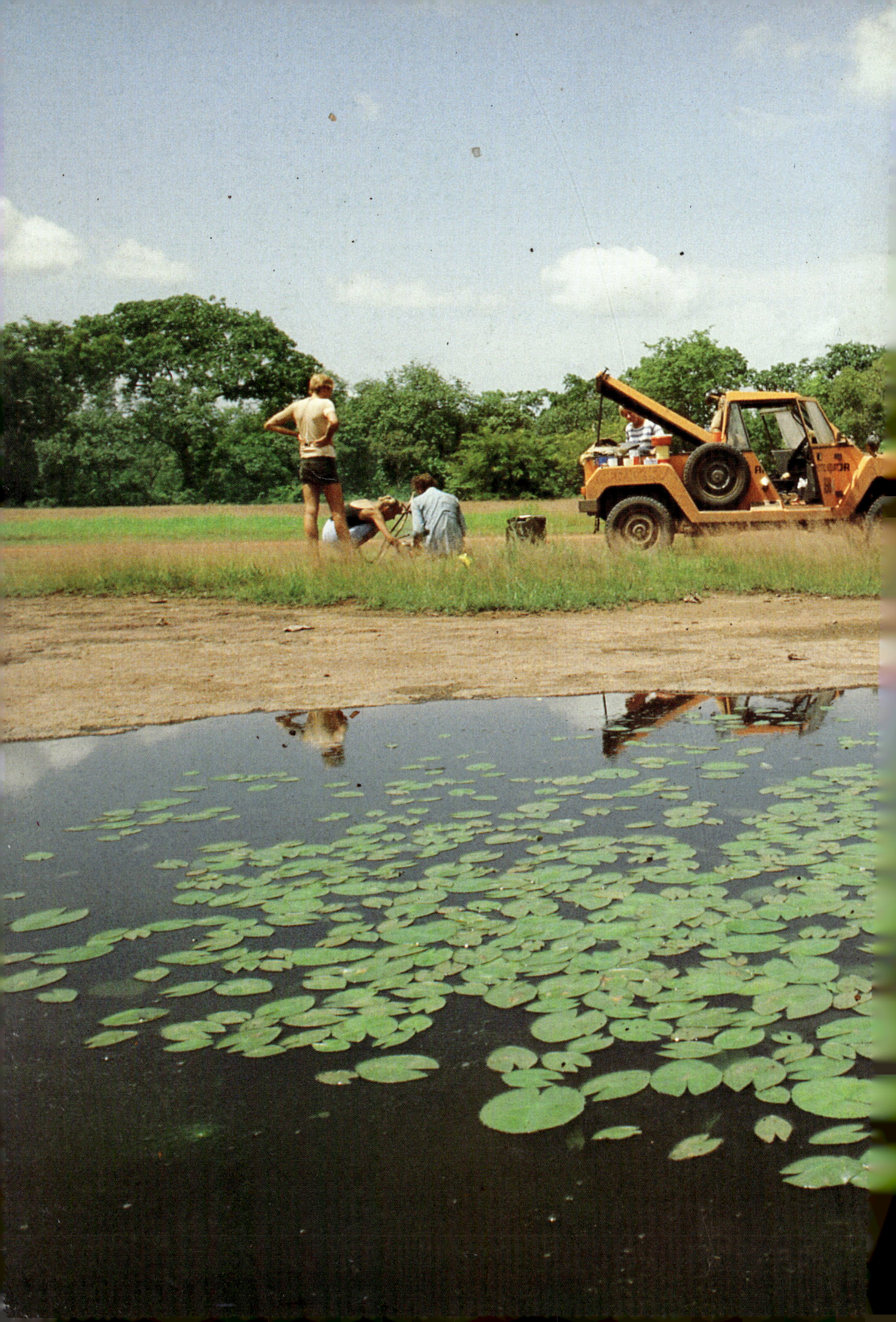

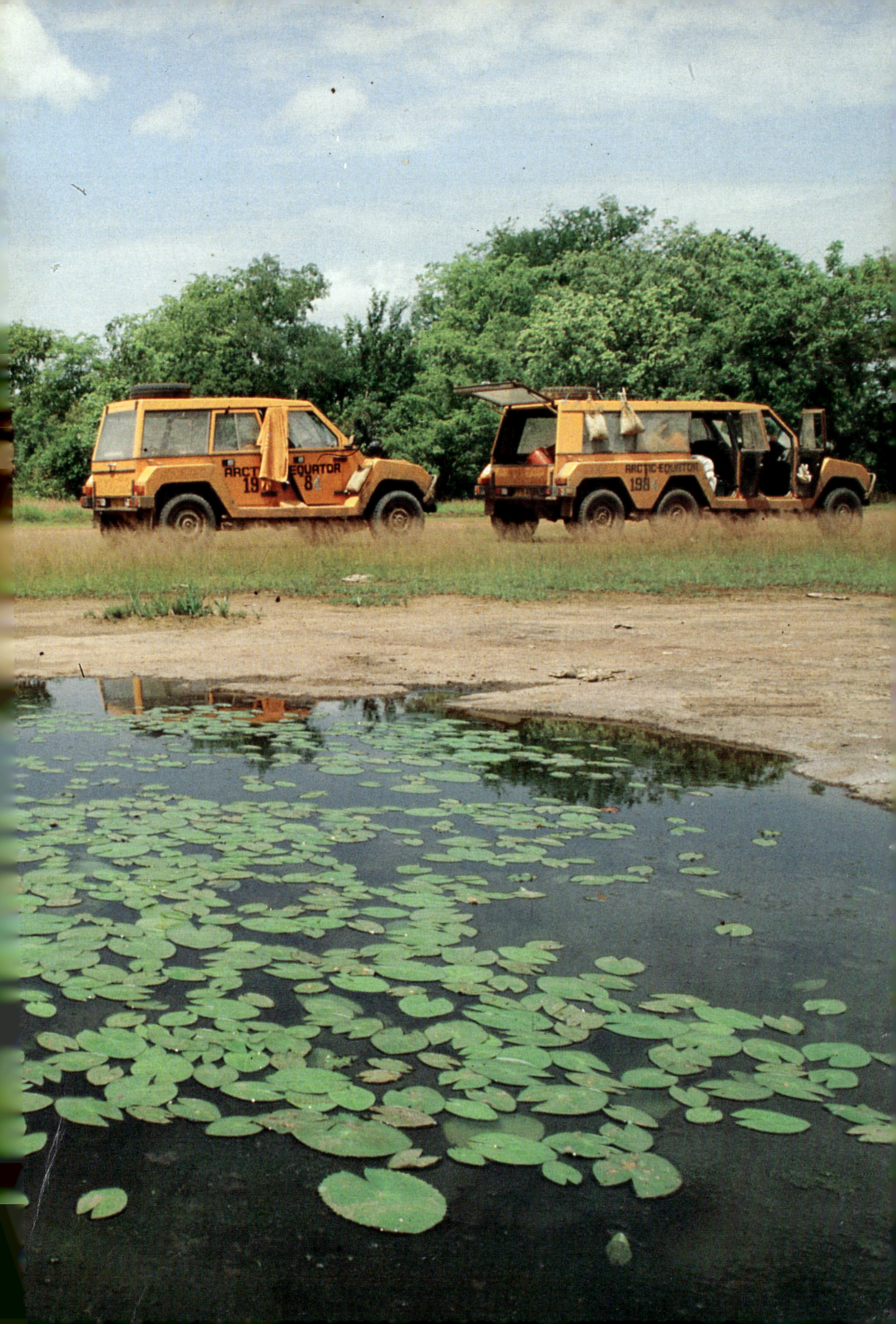
ARCTIC-EQUATOR
1984

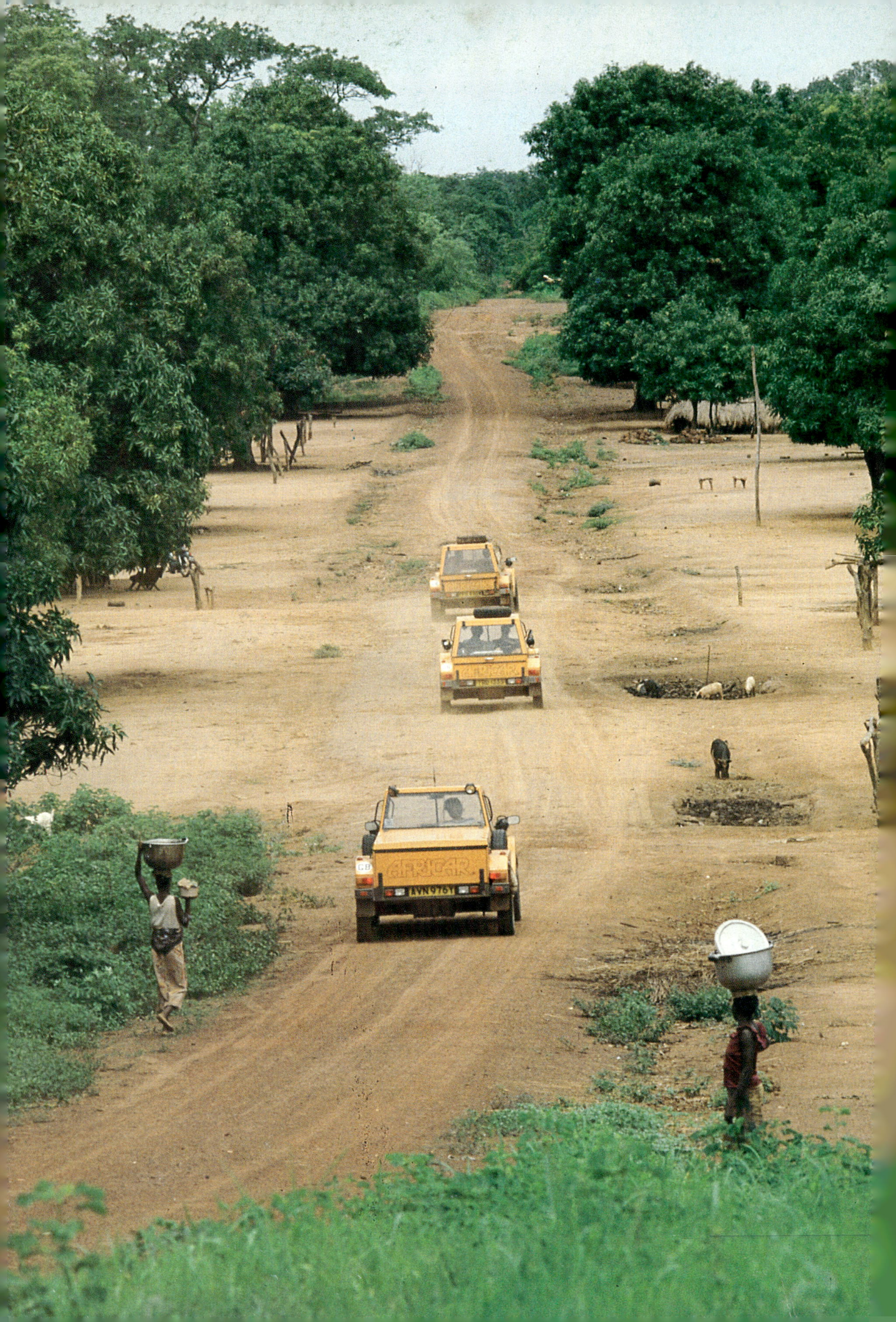

AFRICAR
GB

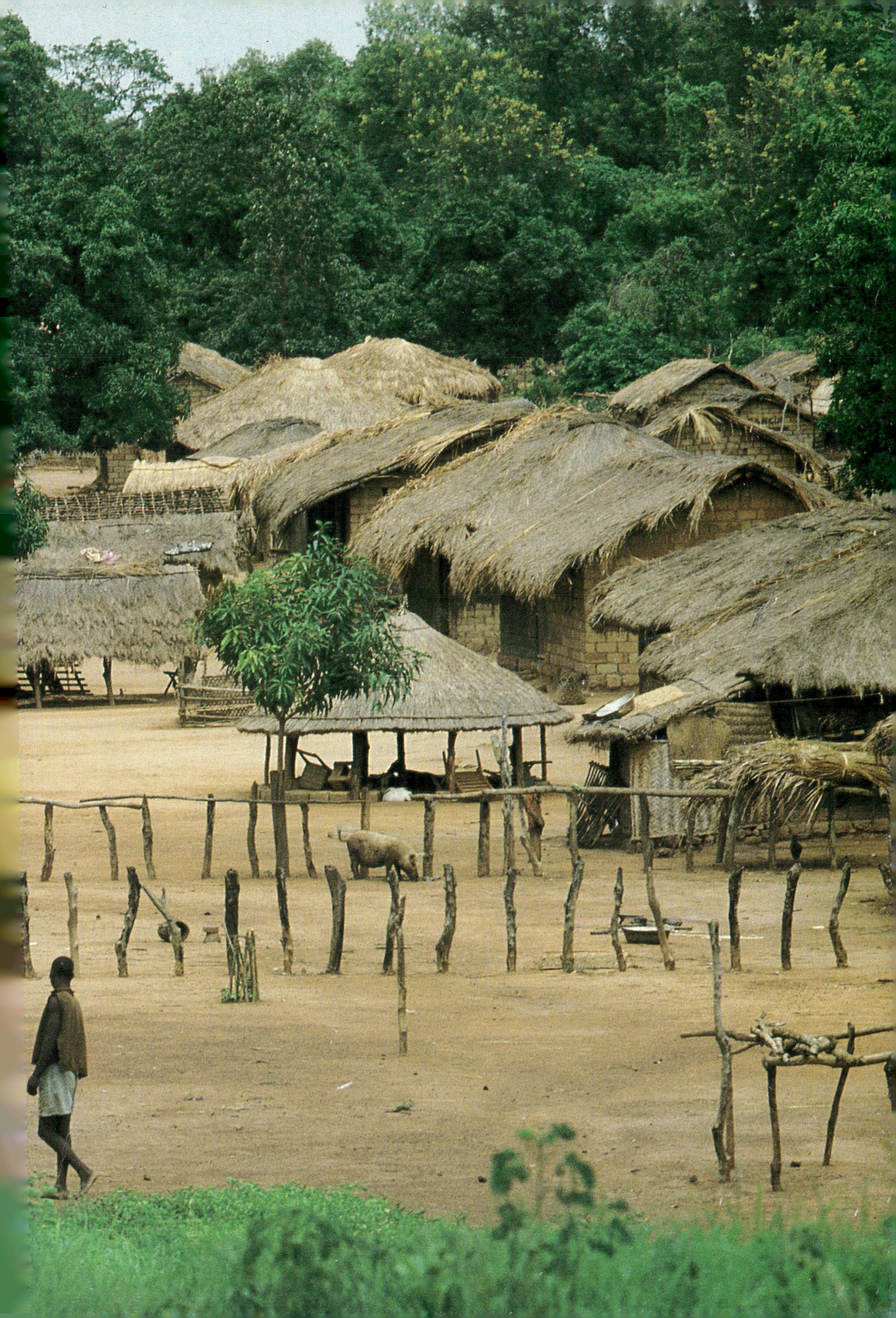

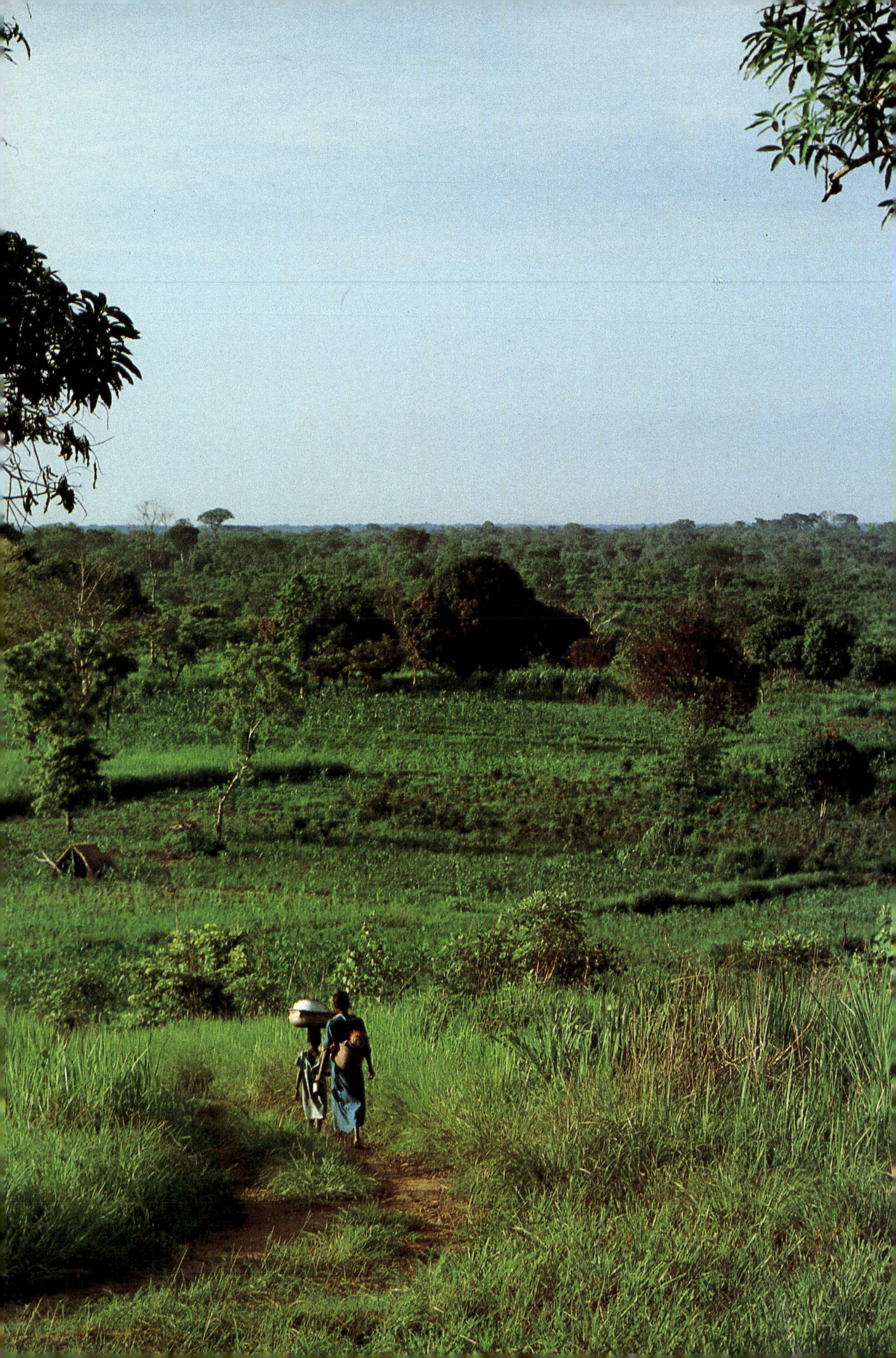

◁ The view from the Baptist Mission at Sibut in the Central African Republic

△ The Baptist Mission

earnest.

GENE *(head of the Mission):* 'I arrived here by the means of a kerosene lantern and a mud hut, when my folks first came out in September 1921 and I was born. I was here until I turned fifteen, left for ten years, then returned as a missionary and I've stayed ever since.

'The first vehicles I can remember out here during the early days were wood burning, used charcoal to make gas, then ran up and down the road.

'This used to be the economic centre of the country. Everything that went to Chad and anywhere in this country came via Sibut by canoe before being distributed.

'It was only when they came in the 1940s with big articulated trucks that maintenance became a problem. Until the roads are raised up above the level of the land, the wet season will remain a terrible time to get around in.'

VERN *(a missionary):* 'I think weight is one of the most important factors for vehicles in these conditions. If they are too heavy you can get into trouble by beating the body to pieces, you also get excess weight on the roads. In mud it's no good at all.'

GENE: 'You have to have something that doesn't beat the driver to pieces. Some of the newer vehicles like the Fords with twin front axles give a more supple ride.'

VERN: 'Height is important but even more so is ground clearance. If you have an oil pan, a tie rod or a steering rod down low, then your clearance height becomes even more critical.

'The other problem is that people have to

▵
In the equatorial part of Africa food, especially fruit, was plentiful.

learn how to drive out here, too. If you know your vehicle is fairly damage proof in the centre you aim to hit the bumps in the centre. If you know it's more to the edge then you drive more to the edge.

'We are finding that independent suspension on the wheels helps a lot. You are going to save by as much independent suspension on the wheels as you can get. And that's part of our problem with the American pick–ups, the twist on the frame tends to break up the bodywork.

'Power is not a factor at all as far as I am concerned. You can get by with a little or a

Despite the start of the rains, the laterite roads of the Central African Republic were ▷ mostly good and fast

lot. Water or air–cooling doesn't make a lot of difference to us here either, but it might in a more remote area. We're impressed with front–wheel drive in dirt and mud.

GENE: 'The vehicles are actually getting worse. The body metal is becoming so thin, some American ones have inner wings made out of plastic.'

VERN: 'You find very few Chevrolets out here and the reason for that is within 5,000 miles on these roads the whole front end of a Chevrolet will just disintegrate.

Yet the 1950s Chevy pick–up was a legend in Africa.

VERN: '50's okay, but don't go up into the 60s and get one. The beds on American pick–ups are the worst ones, they are light-weight and crack up. We take them off and reinforce the bottoms, the tailgate and the front support where they are mounted on

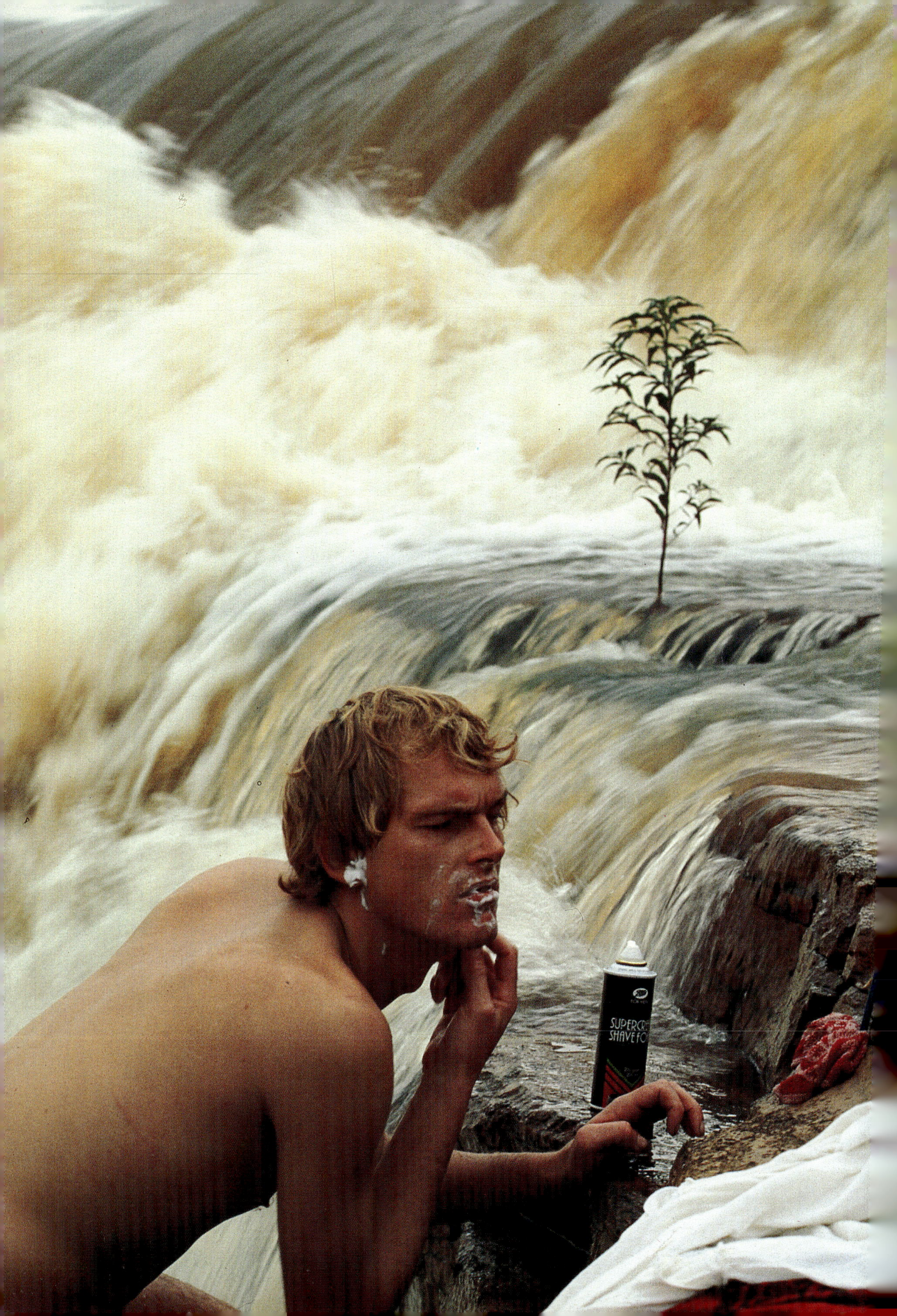

the frame, that all has to be re-welded. Otherwise 15,000 miles and it's rattling like a tin can.'

When we arrived at the Mission, the grass cutter wasn't working nor were any of the twenty or so relatively recent American four-wheel drive pick-ups and wagons that littered the compound.

But we also had our own problems. A design policy was starting to catch up with us. We had decided to build all prototype suspension parts as light as possible in the hope that fatigue cracks would eventually indicate just where the complex dynamic forces were acting.

After 25,000 kilometres the policy had started to work, although we found the main problem was due to incomplete original welds. Quality control in prototypes is as critical as in production.

At the Mission the rains broke. Gradually, from now on road conditions would deteriorate and we would find out if our test conditions in England, a country that is no stranger to mud and rain, had been

◁ **Camped besides the glorious waterfall at Kembe in the Central African Republic**

At this stage thunder was continuous and cloud build-ups were always ahead of us
▽

△
Heading for the Ubangui River and the crossing to Zaire

adequate and if the results were truly applicable to Africa.

We don't really expect any major problems until we reach Zaire, but the frontier is only another 400 kilometres away.

At Kembe we washed ourselves, our clothes and the cars in the waterfall, and we were able to buy avocados, pineapples and mangos. So, cleaned and polished, well fed and relaxed with the cars in good shape, we drive off on the last good laterite road before Bangassou. There, if the ferry is working, we have our last chance to cross the Ubangui River to Zaire.

If the ferry is out of service we can't go forward, as the way to the east is blocked by the civil war in Sudan. We will have to back–track 1,000 kilometres to Bangui and try there.

Without telephones or any other forms of communication, the only way to find out is to go and see.

7
EQUATOR

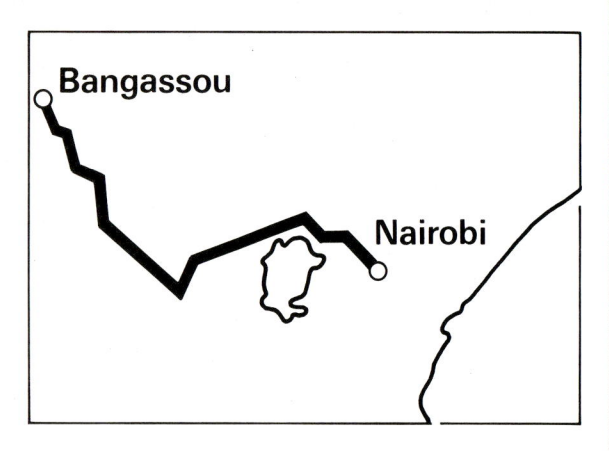

Zaire occupies the part of Africa that used to be known as the Congo Basin. Situated right on the Equator, it is a low lying area of swamp and jungle separated by a myriad of rivers which are swelled annually by fierce equatorial rains.

Building roads in Zaire is a thankless task. There are thousands of streams, tributaries and main rivers to cross with little between them in the way of solid foundation. Laterite surfaced roads of the colonial period only survived through intensive maintenance. Today, in the more remote areas of Zaire, these roads have lost their protective surface and simply disintegrate when the rains arrive.

It is not unknown in the rainy season for a truck to get stuck, blocking a road closed in on either side by a solid wall of jungle, and causing a traffic jam in which vehicles are trapped for weeks, sometimes months at a time.

Overland travellers try to avoid the rainy season in Zaire. The holes in the roads can be literally truck–sized and bridges are swept away. It is a claustrophobic place in which to be stuck, as vision is reduced to a few metres on either side by the impenetrable jungle and to a few hundred metres down the track. There are hundreds of kilometres without supplies and no communications. Many of the old towns or trading centres were abandoned during the civil war of the early Sixties and never re–inhabited. The population still seems nervous from those violent times.

Now the rains have caught up with us and we will find out if the AFRICARS can cope with Zaire mud as well as with desert rocks, sand and corrugations.

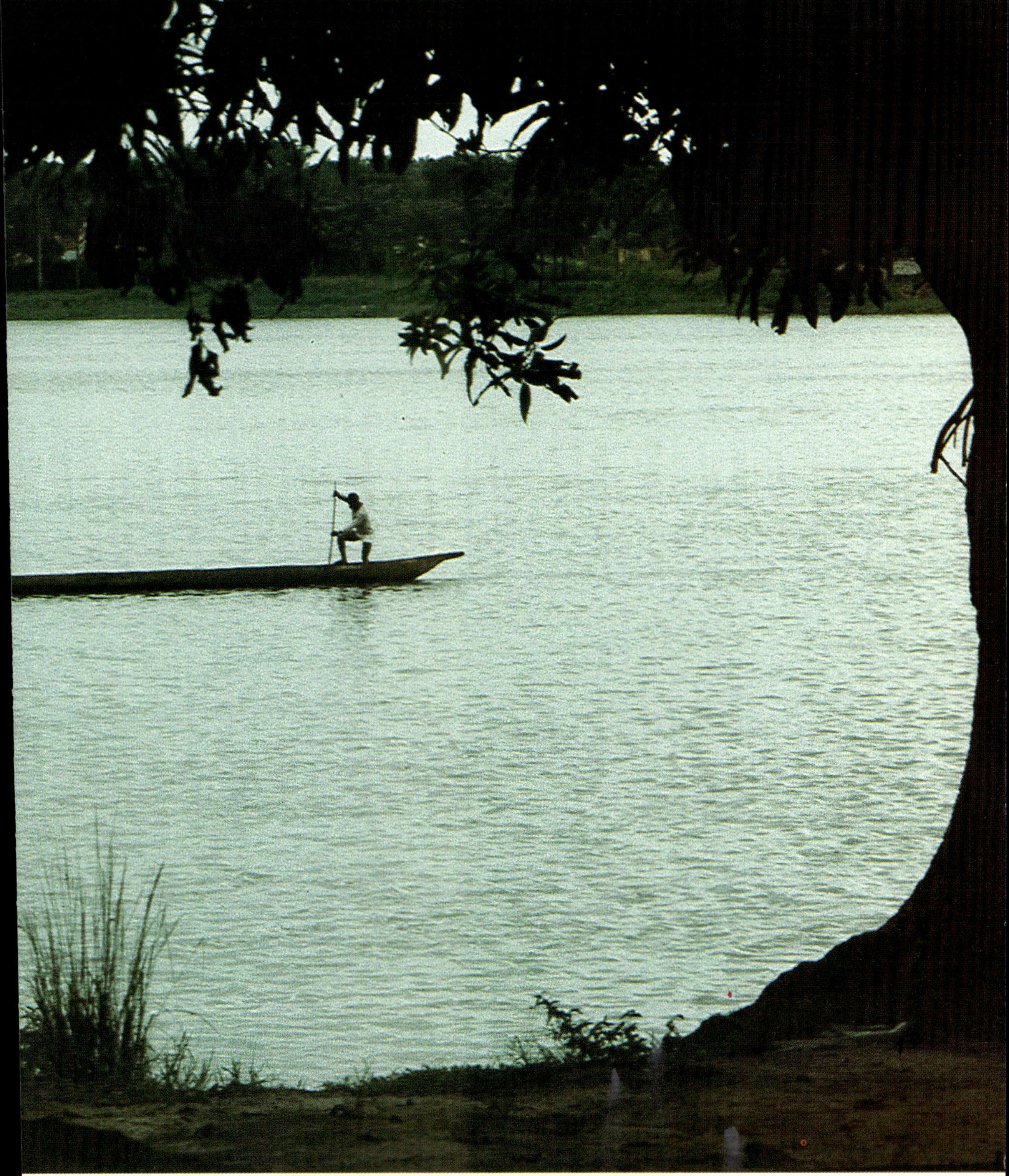

In Equatorial Africa, the area that used to be called the Congo Basin, there are thousands of rivers and river tributaries. Almost every main road is served by diesel powered ferries, large enough to carry trucks as well as cars.

If the roads are passable we will drive south to Kisangani, capital of Zaire's Eastern Province and then head for the Ruwenzori Mountains, Uganda then Kenya.

The problem is that first you have to have enough battery power to start the big diesel engines, then you have to have enough diesel fuel for the crossing and lastly you have to go and catch your ferry - which will, of course, be on the opposite side of the river.

The enshrined laws of the universe operate as inconveniently in Africa as they do in the rest of the world.

Just before reaching Bangassou we had met John, a Dutch carpenter, travelling on his own in a Land Rover, from Cairo to East Africa. Forced to take the long way round from Khartoum, because of the civil war in Southern Sudan, John is now short of money, food and fuel. Apprehensive about the roads in Zaire, he has welcomed a chance to travel with us to Uganda.

As the ferry reached the Zaire side of the river the ratchet slipped on the loading ramp. Camera, cameraman, recorder and sound recordist plunged deep into the muddy and swollen river.

If we'd been using video that would have been the end of the film. But the French film camera and Swiss recorder proved to be well made and well sealed. Apart from a slight mistiness in the lens for a few days, they were back in action within a couple of hours.

The main road south from the second largest town of the Central African Republic to the capital of Zaire's Eastern Province is treacherous for any vehicle during the rainy season.

After the wide open spaces of the desert and the soft airiness of Chad, there is something a little sinister about Zaire. The jungle is closing in, blocking our view in every direction, the sky itself is disappearing.

We pass through settlements boarded up and secured with rusting padlocks. Abandoned in the civil war of the early 1960s and never reopened.

△
The ferry at Bangassou across the Ubangui River to Zaire. Here we needed to supply batteries to start the ferry, and diesel to run it

Immediately in Zaire the road deteriorated ▷

164

Those were violent days when Belgium, Britain, America, South Africa and a clutch of multi-national corporations sought to separate the mineral rich Southern Province of Katanga from what had been the Belgian Congo.

A situation finally brought under control by United Nations forces but not before anarchy and violence had reached appalling proportions.

I remember that time and to me our claustrophobic and sodden jungle camps seem full of ghosts – Dag Hammarskjold, Secretary General of the United Nations, 'killed' in an unexplained plane crash and the hundreds of thousands who died for nothing more than greed and a misjudged adventure planned tens of thousands of miles away.

The jungle is now so dense that we have had to travel 20 miles to find a tiny clearing.

Instead of desert sand it is equatorial mud that hinders the repair of another of our temporarily borrowed and abused gearboxes.

The gearbox, split open at one end, was stuck back together with black mastic and held by a steel strap, a repair almost as good as the hardwood piston in a six cylinder truck engine, that we heard about in Zaire. Ours lasted till Nairobi. I can't vouch for the life of the truck engine.

Visibility is everything on these roads, almost as important as ground clearance.

The roads are rarely maintained, and then only to suit truck traffic. Wheels on a vehicle for Africa need to be set wide apart to match the track of at least the inner wheels of the average truck.

This is especially the case with many of the log bridges. A truck or an AFRICAR can drive straight across. A narrower vehicle often has to negotiate the crossing with the delicate balance of a tight-rope walker.

Tyres scrabble for grip on the round and slippery sides of the logs as yet another bridge is successfully negotiated.

The Land Rover first became stuck in the middle of the road late at night. We pulled it out with the wagon. The next day we had to back track because a ferry had sunk and John stuck twice in the same day.

The problem is rather obvious. A solid

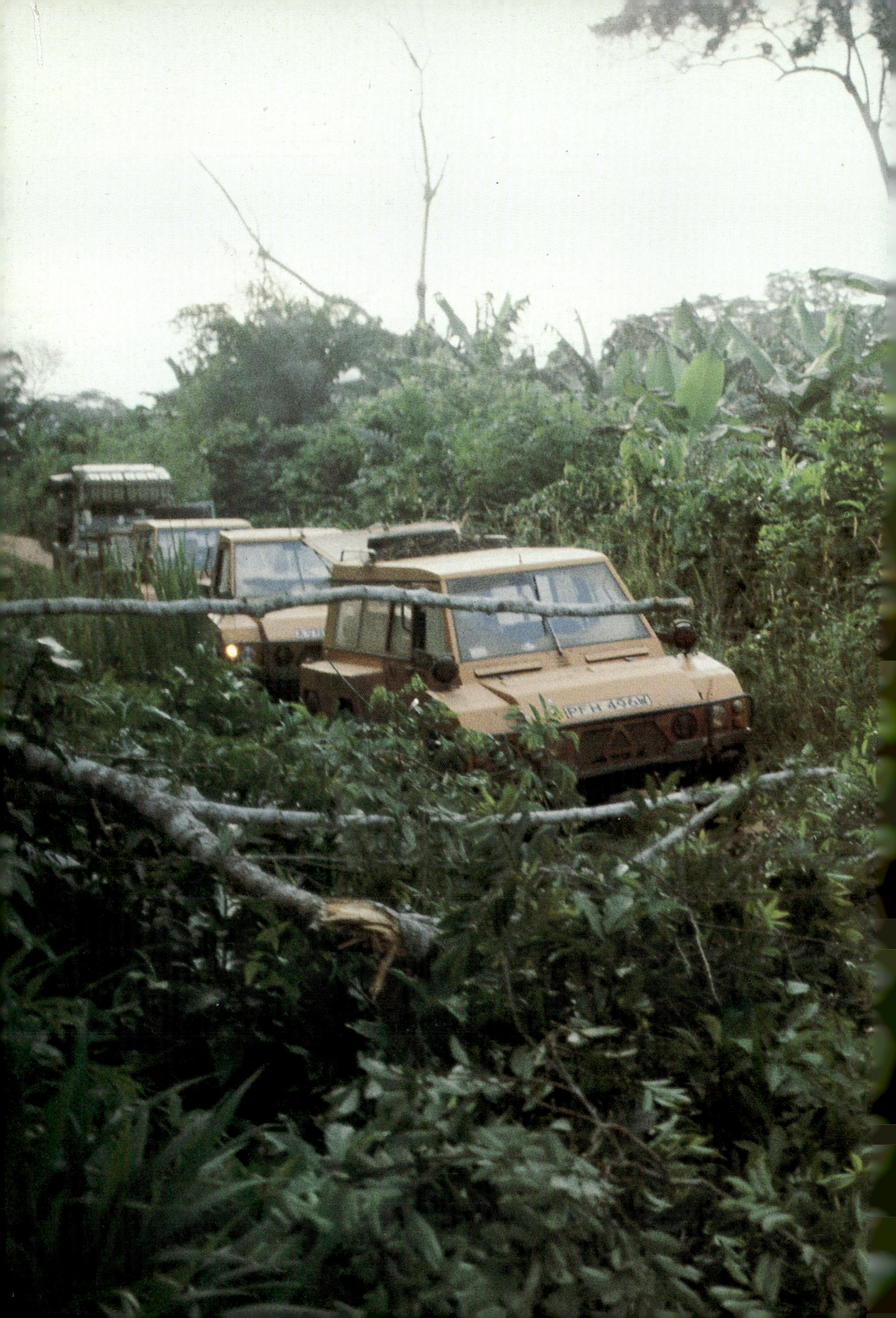

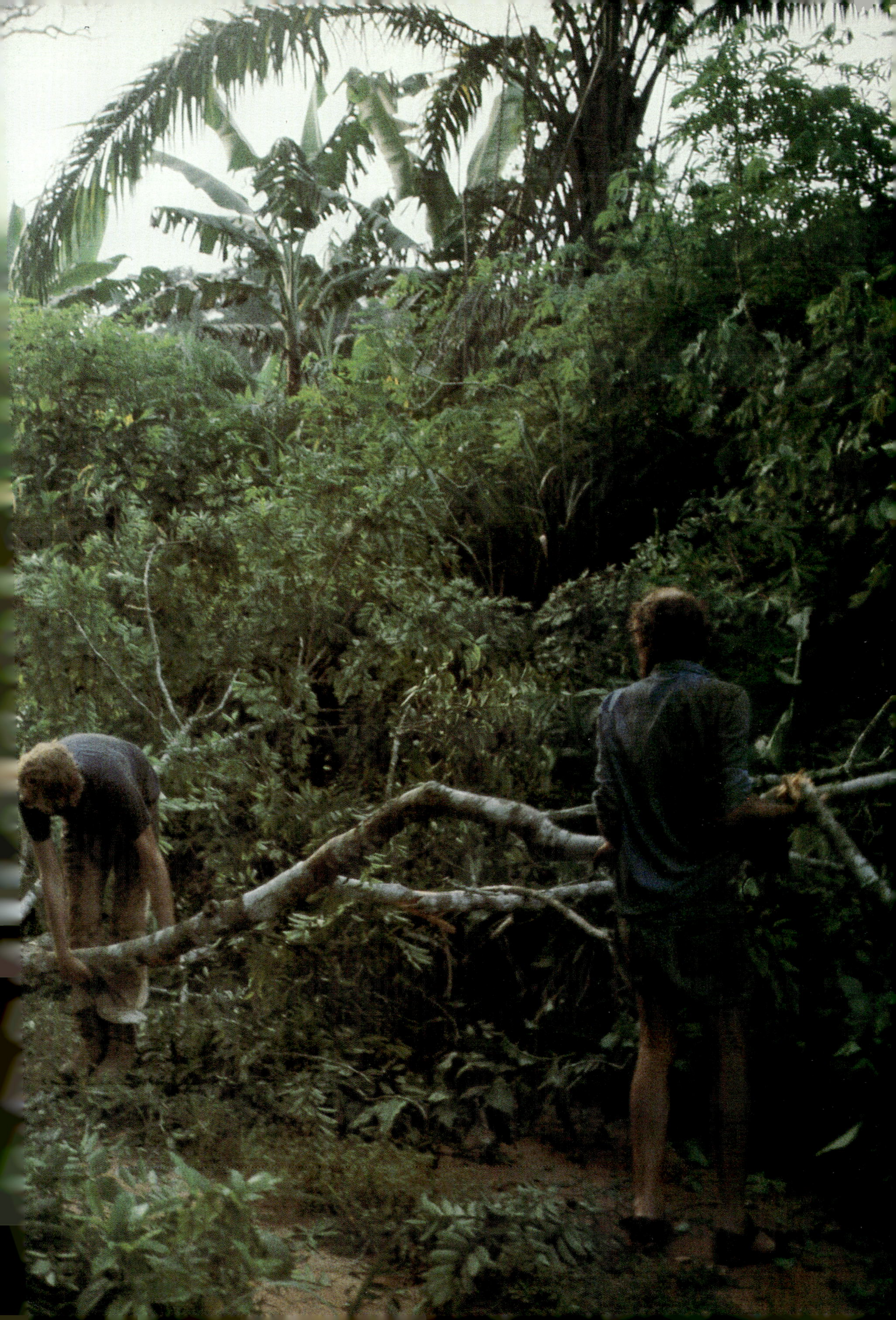

▵
We met John from Holland travelling on his own in a Land Rover

The Land Rover had difficulty with the slippery log bridges, partly because it was too narrow and partly because of the visibility problem ▷

beam axle with a differential in the middle tends to have about 8 to 9 inches ground clearance or 200 to 225 millimetres. Once this is used up the axles simply sit on the road and the wheels are off the ground. No one is going anywhere fast in Africa. It happens all the time.

The problem with the axles is made worse by the long and relatively low overhang at the back, one of the main reasons for building long-chassis AFRICARS with 6 wheels and little overhang.

A car made or sold in Africa must be able to negotiate normal African roads. There is no excuse whatsoever for a car, let alone a four-wheel drive all-terrain vehicle getting stuck on a main highway.

Testing in England for African roads had been equally wet and muddy but much colder.

The characteristic of the AFRICAR established with the early prototypes is quite different to conventional four-wheel drives. The AFRICARS have over 12 inches or 310 millimetres of clearance and the smooth underside of the chassis can float and slide on the mud with the independently sprung wheels free to reach down and find traction.

The AFRICARS, while being tested on Salisbury Plain, managed well in two-wheel drive in 18 inch or 450 millimetre deep ruts.

At that stage the only problem left to solve was how to keep the mud and water out of the engine and the electrical system.

It was this type of central African earth road that we really designed the AFRICAR for in the first place. When we considered the width between the wheels, the ground clearance, the general stability of the vehicle and its traction, it was this kind of road that we had in mind.

Immediately after the rain the road is very difficult and dangerous to drive on because the potholes fill up to the top with water and you have no idea what's underneath. It might just be a few centimetres or it might be a metre or

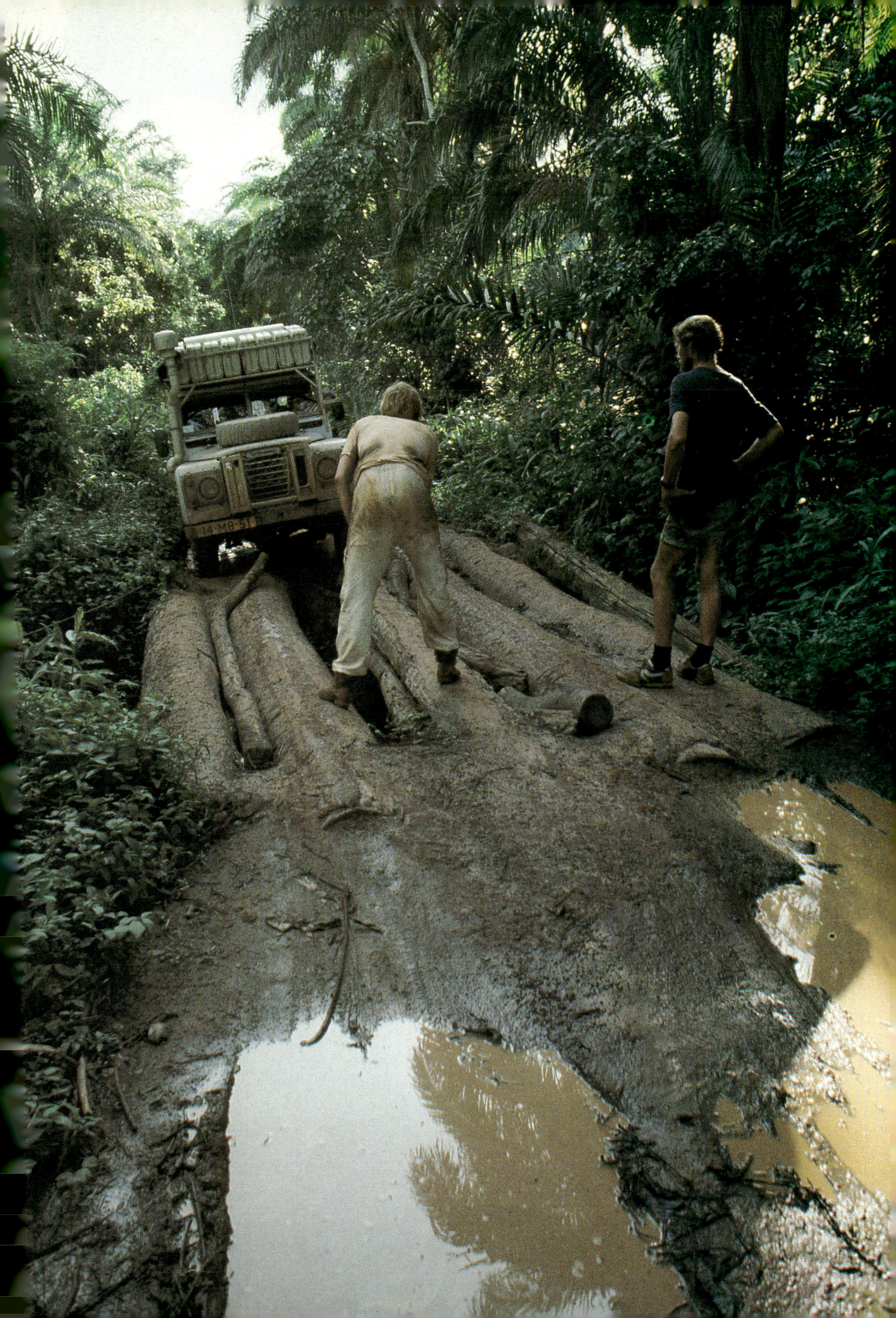

two. Some holes on these roads can swallow whole trucks, let alone cars.

For people who habitually travel these roads there is nothing unusual about getting stuck - frequently.

What was unusual was the fact that, in two-wheel drive only, none of the AFRICARS got stuck once on the road at any time during the trip. Only in the desert with no road at all, like everybody else, we needed the occasional push out of soft sand.

This is the second time in two days that you've been pulled out of the mud. What's actually happening when you get stuck?

JOHN: Well the axle touches the ground. That means we're finished.

You've got no drive on either wheel? Has it happened often before?

JOHN: Yes many many times.

But you manage to get out on your own? How do you do it?

JOHN: Normally I have a passenger that can help me.

This hole over here is the one you got stuck in the other night. We're going to try and bring the two AFRICARS through and see whether they can run straight through the middle, like you did, but without getting stuck.

With the total lack of concern to which we had become accustomed, the AFRICARS cruised through the hole, and just to prove a point, we reversed them back through it again.

When the sun broke through the rain clouds, Zaire became a warmer and friendlier place. In the Eastern Province, bordering Uganda, the landscape changed yet again, becoming open and less threatening.

Last night we broke our last remaining rear suspension unit. We broke one previously on the pick-up, which is not surprising because the bump and droop stops are not in the right place.

This has been and still is the biggest crisis that we've had on the whole trip. In theory, you can't repair these units.

When I first talked to Tony Best about using these I said to him, the problem with anything like Hydragas is that you've got a unit which you have to replace if anything goes wrong,

△ Everywhere you go in Zaire there are ferries, some motorised, some worked by hand. But they are free and usually operating. We were never held up by them

Camping in Zaire was wet, muddy and rather miserable the road became too bad for the Land Rover. Always the same problem, insufficient ground clearance under the axles ▷

whereas with a leaf spring you can weld it up, or take a piece of metal and make one. I mean it's not like having a leaf spring that a blacksmith can put right.

TONY BEST (in UK): No, but the intention is that they are made and sealed for life.

You know I always dread that phrase, I mean 'sealed for life' is the ultimate killer.

How do I explain to people the real problem? That is you've got something like this and you just have to take it out. throw it away and find another one under the nearest baobab tree, which is most unlikely.

TONY BEST: Well we try and design them so that in fact they last the life of the vehicle.

The crack is right in the centre. We've taken some Tip-Top rubber solution, the universal tyre repair solution you find everywhere in Africa, and some old inner tube. We cut patches and glued them in the centre, so that when the piston goes in, it goes into the new patch and hopefully seals the unit.

A typical African repair using the available materials. We've sealed up the end of this and last night we put 500 pounds per square inch pressure in it.

THOMAS: You can only put a patch in the middle of the unit, where the piston goes in because in the sides you can't reach it. So it is only possible to repair it in the middle. If it works?

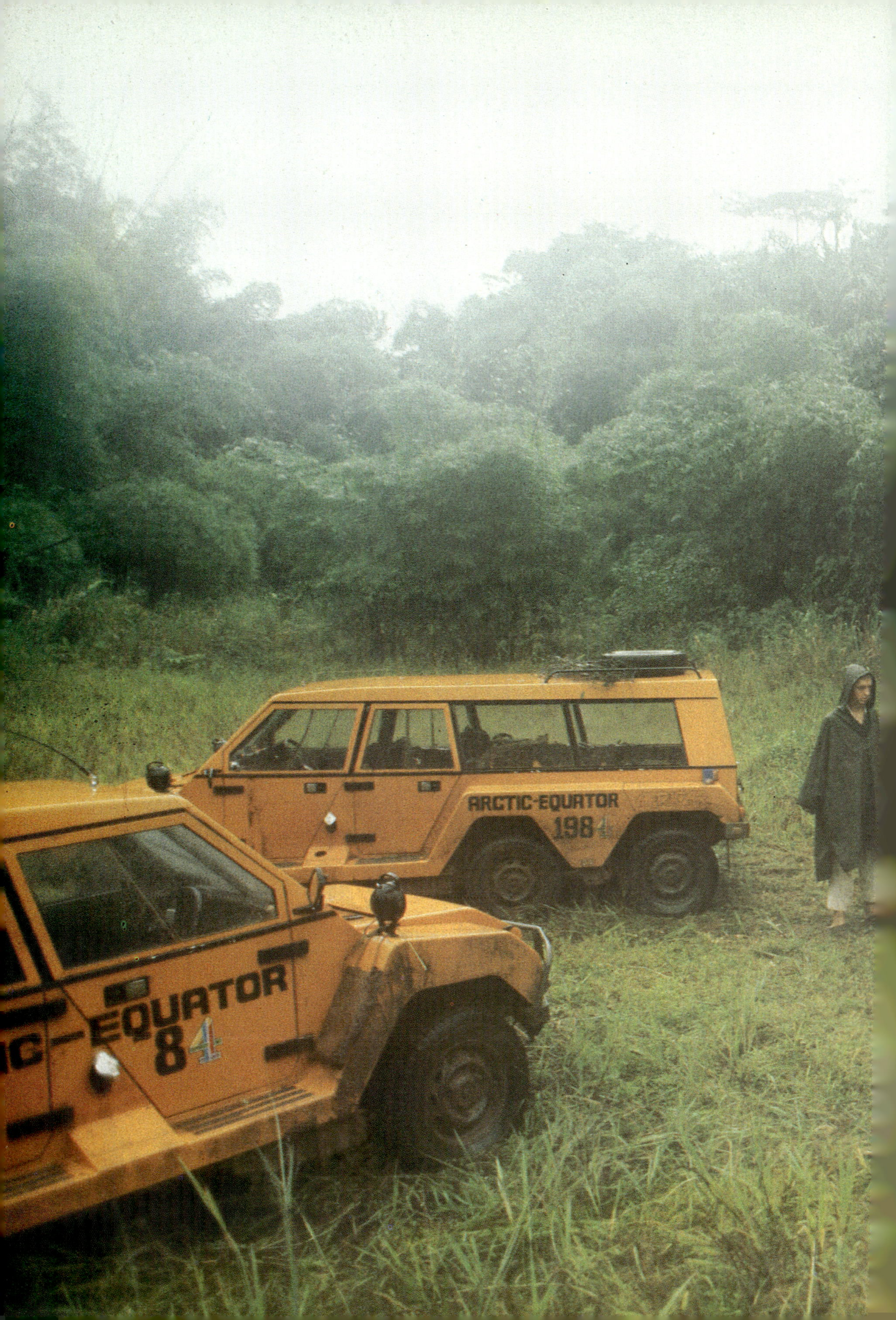
ARCTIC-EQUATOR
1984
EQUATOR
84

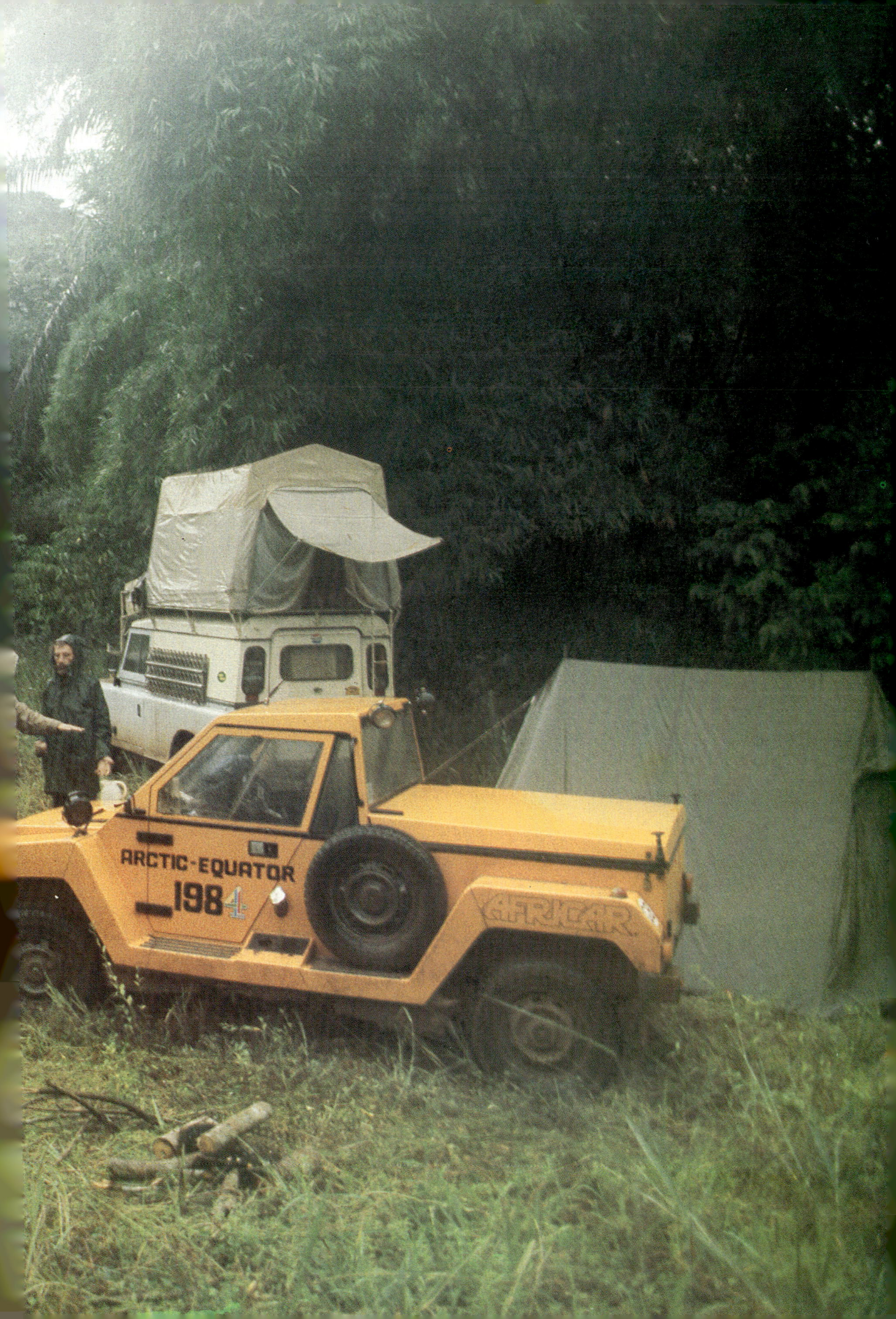
ARCTIC-EQUATOR
1984
AFRICAR

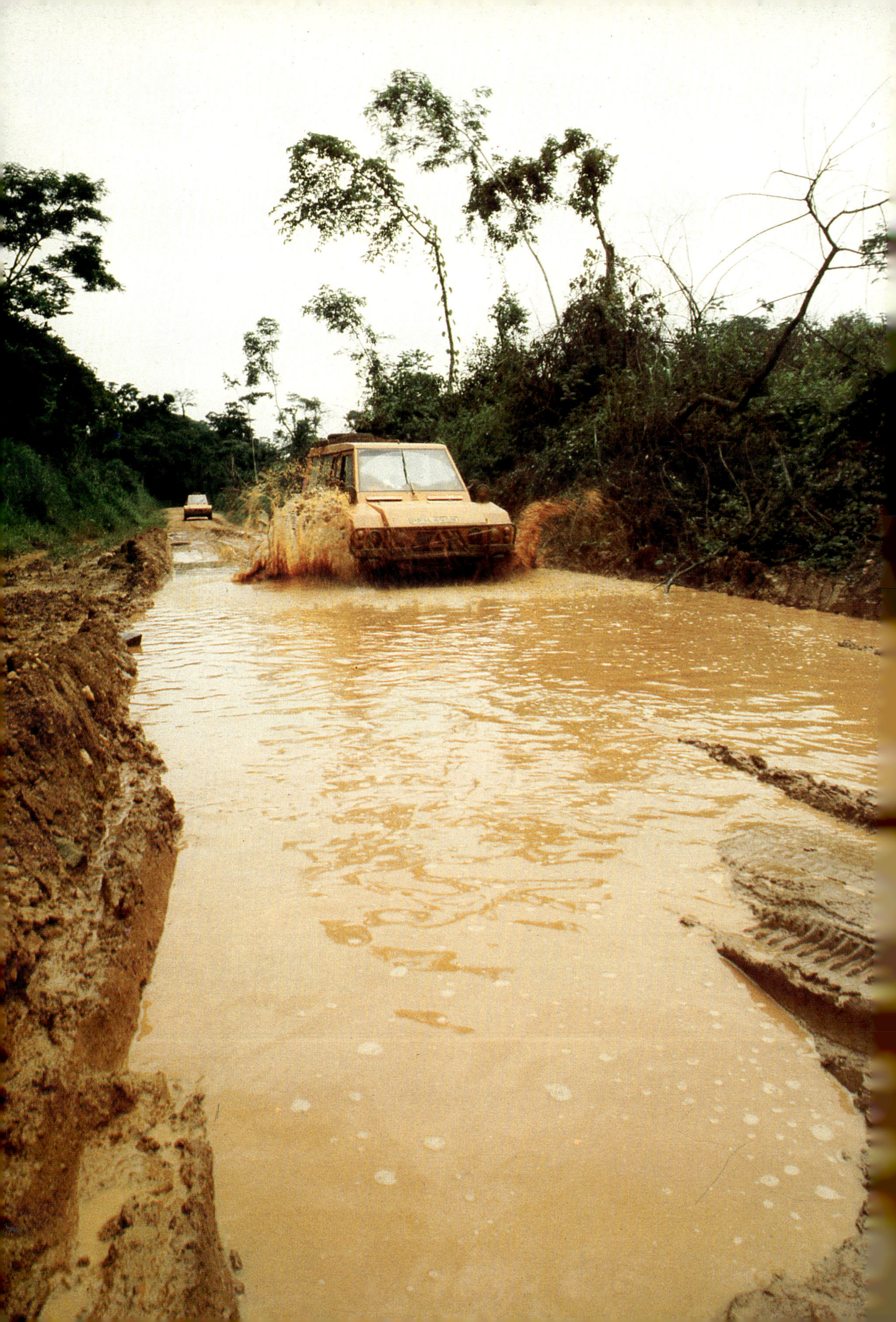

◁ A higher ground clearance on the AFRICARS
allowed us to negotiate all these roads
without a problem

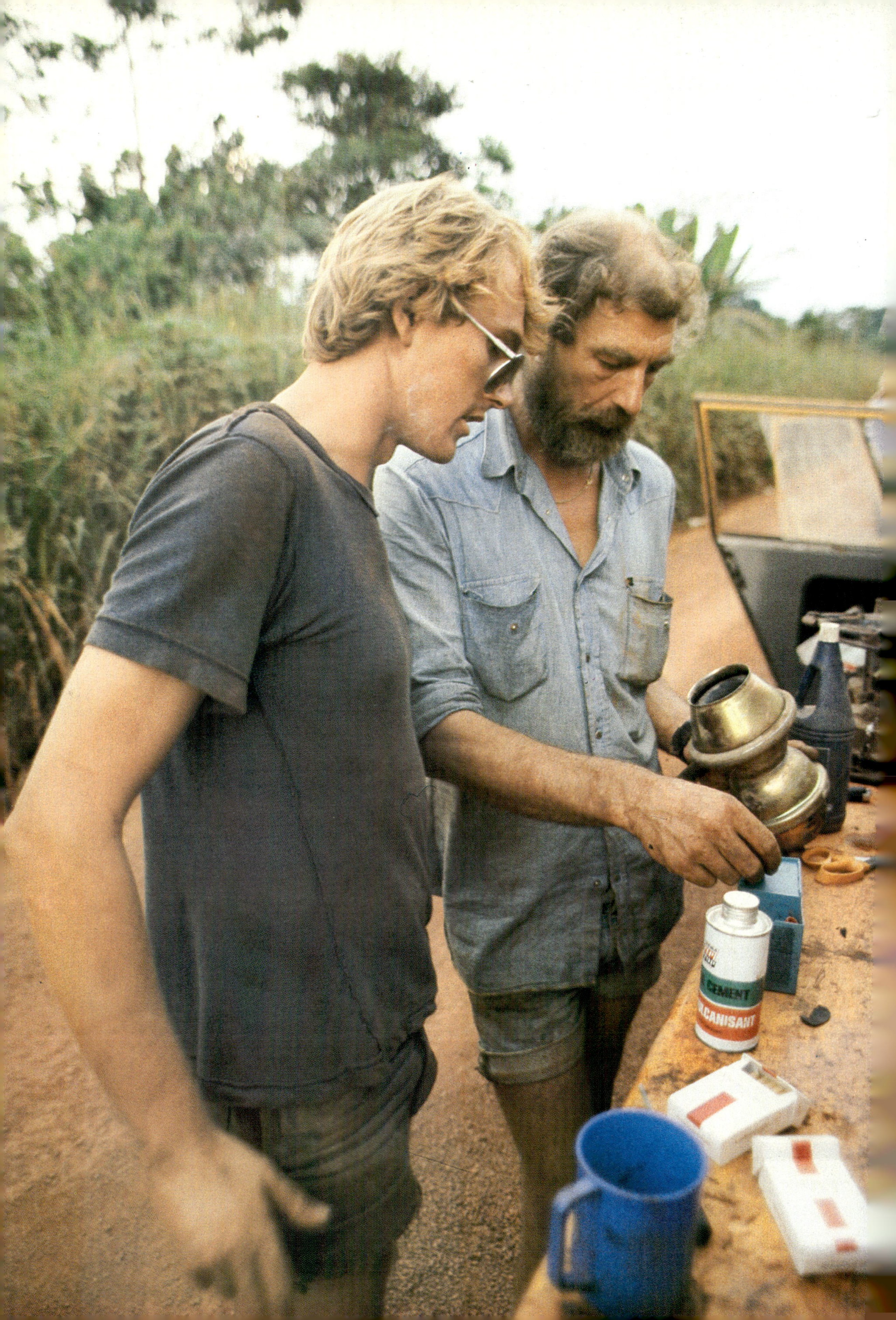
CEMENT

◁ With Thomas and the suspension unit that we repaired with Tip Top puncture repair materials

The jungle is beautiful but oppressive. For ▷ days we saw nothing but the road ahead

△
The wagon gets a push as it tries, successfully to pull the Land Rover off another mudbank

What I've decided to do to discipline our driving is to ask John to run in front with the Land Rover and set a Land Rover pace.

So the Land Rover had its revenge. And Tony Best was proved right. The problem was caused by another prototype quality control slip and not any deficiency in the life of the unit.

The rear suspension pushrod had a machining error that left less than a millimetre of steel to carry up to 6 or 7 tonnes load each time we went over a bad bump.

The Land Rover's revenge didn't last long. Within a hundred kilometres it came to a grinding halt.

THOMAS: What's the problem John?

JOHN: The rubber at the bottom of the shock absorber has gone, which means a lot of shock on the leaf spring. The bolt in the middle of the spring has broken which caused the front wheel to move backwards.

THOMAS: The vibrations made the whole spring move backwards and now he's hammering it back into its position. I personally prefer to repair a spring with Tip Top.

John was very capable and seemed Okay. We were losing about twenty minutes in every hour, by travelling with him in addition to the time spent getting him out of the mud. We now felt obliged by our own

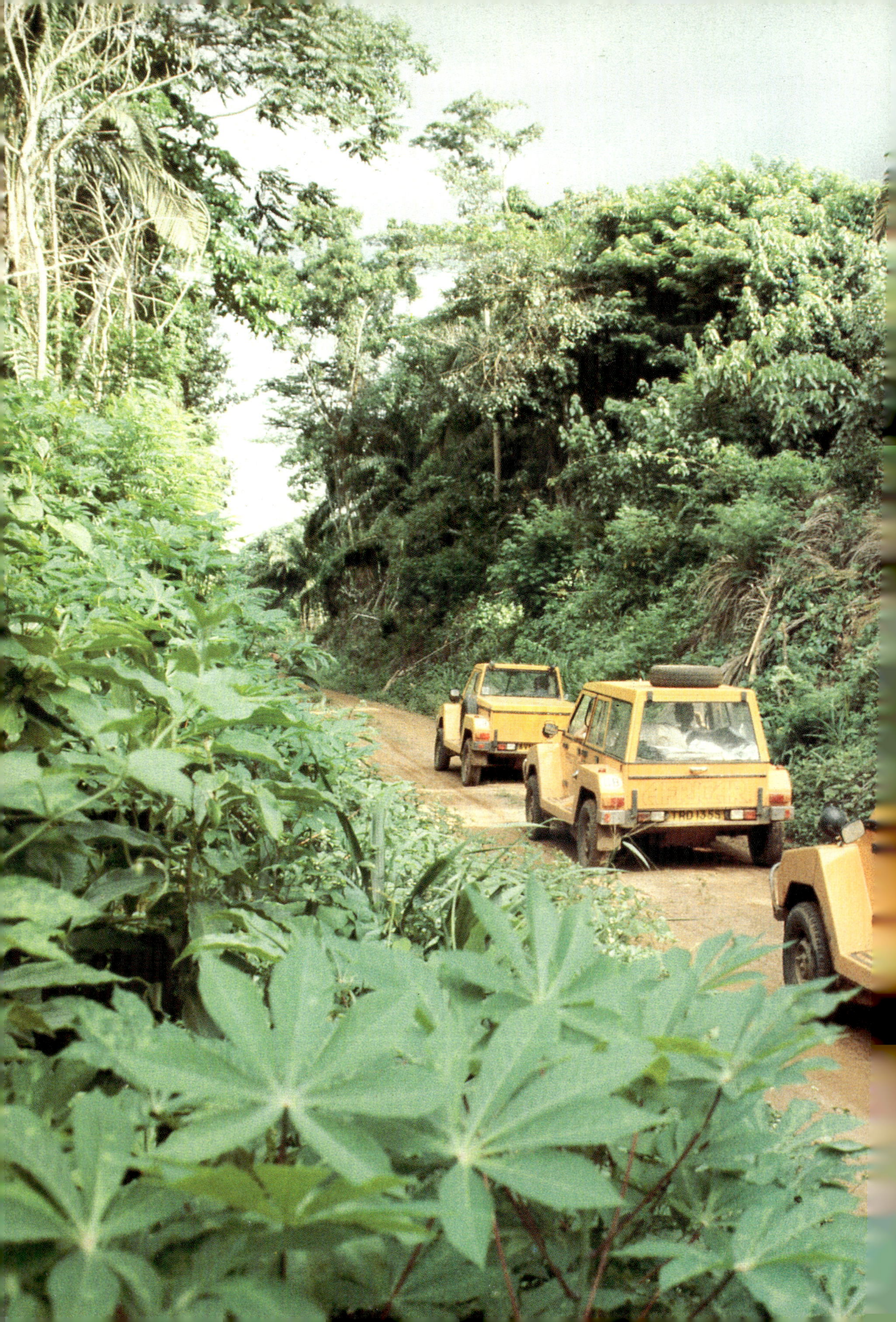

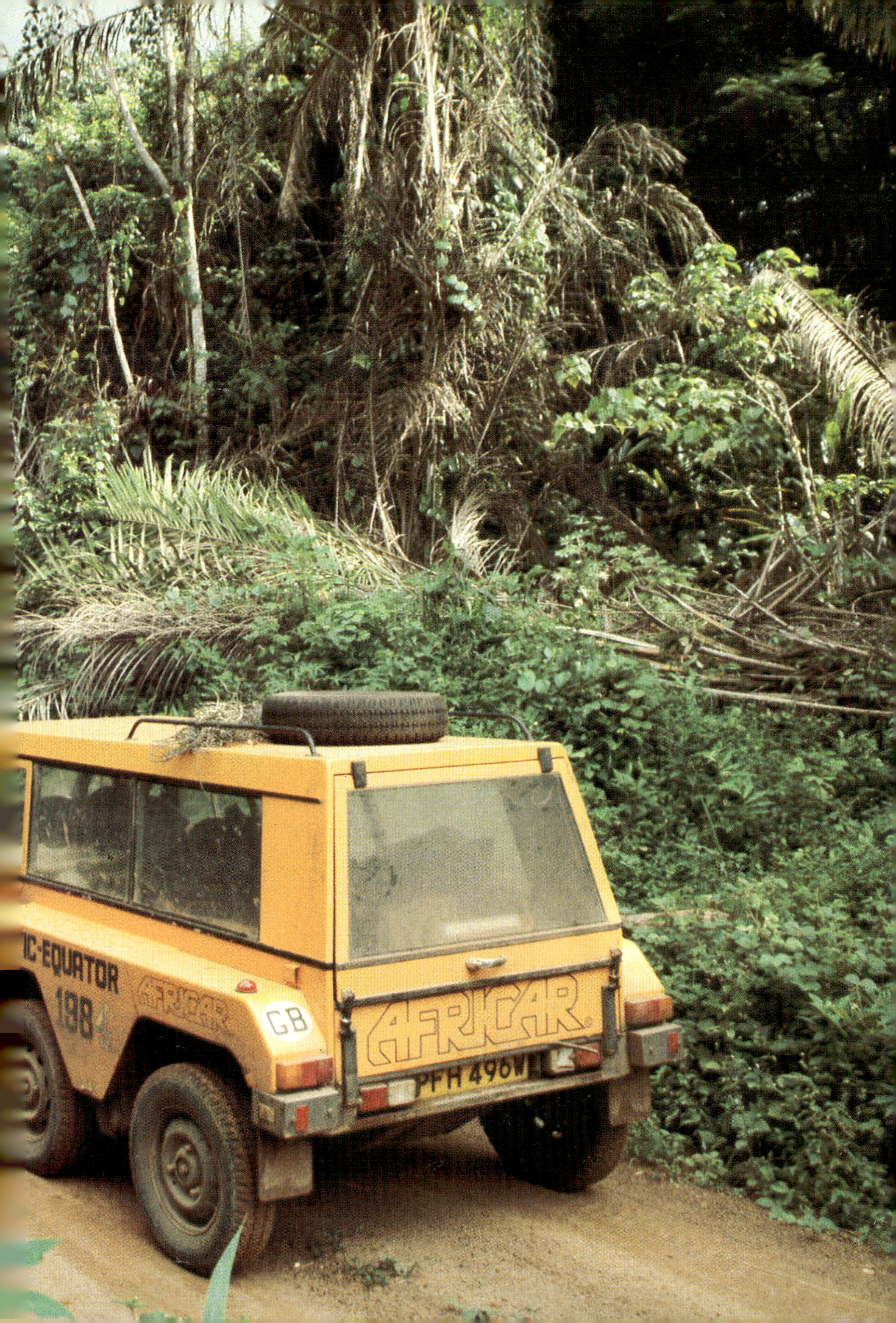
IC-EQUATOR
1984
AFRICAR
GB
AFRICAR
PFH 496W

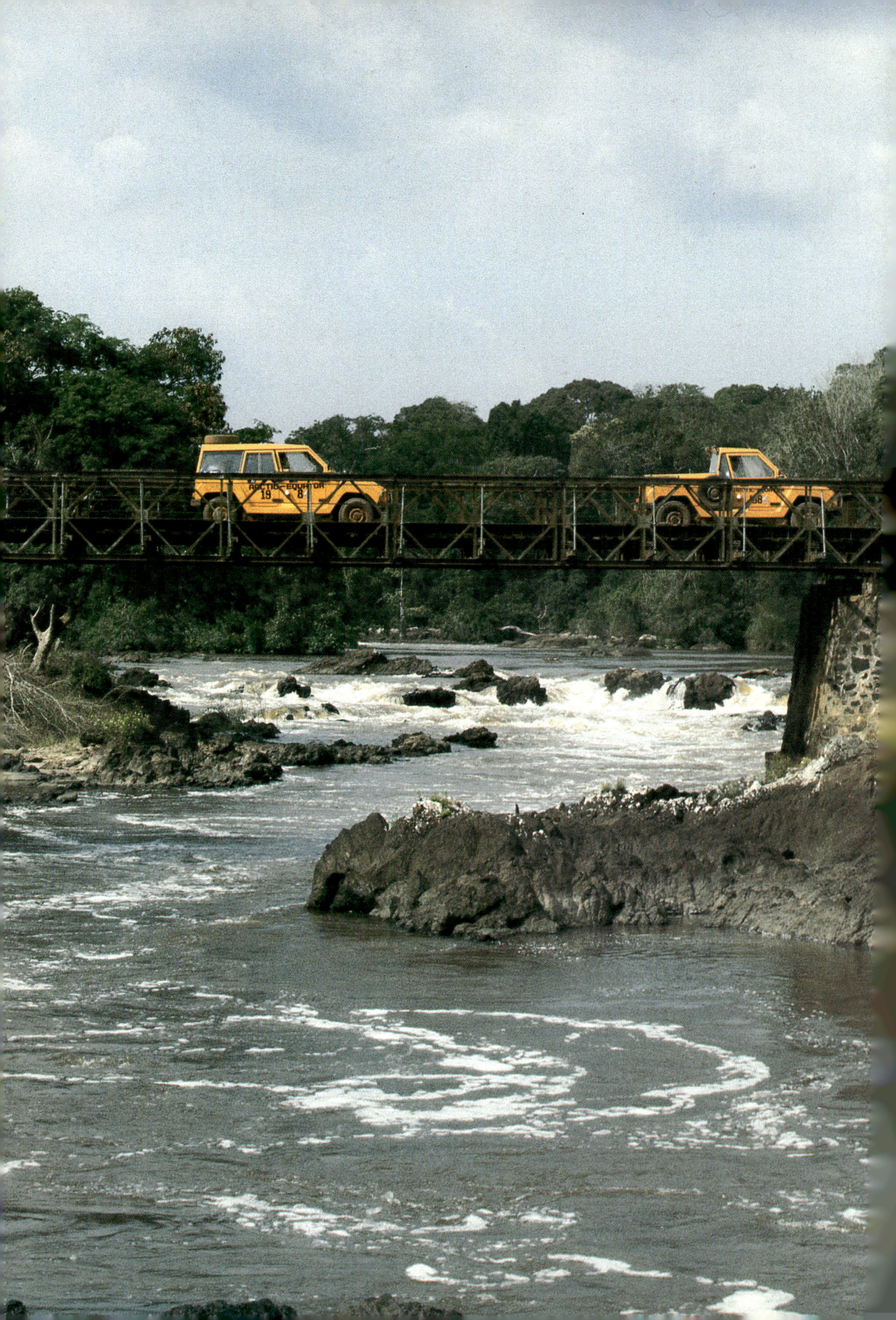

△ A rare okapi, an animal that seems to be a cross between a Zebra, a horse and a giraffe, at the Egulu camp

The Epulu River in Eastern Zaire

The crew at the Equator sign in the Ruwenzori National Park, Uganda ▷▷

commitments to leave him wrestling with his springs and to go on ahead.

To make up time we decided to take a short but doubtful road straight over the Ruwenzori Mountains to Uganda. The surface was atrocious and we were lucky that it didn't rain. This was our first experience of black cotton soil, almost bottomless when wet.

Bridges blown up in the civil war twenty years ago, have only been temporarily repaired.

By this stage of the trip everyone has developed their own particular AFRICAR driving style.

Charles makes a valiant attempt to run over the cameraman.

Carolyn calmly and simply drives over the obstacle. No fuss, no problem.

And Thomas - well Thomas . . .

As we reached the top of the pass through the Ruwenzori Mountains we found, to our amazement, that we were on a truck route. That explained the huge holes in the black cotton soil further back.

Scraping our way past on the portion of the road left to us, we silently wished them luck with the bridges.

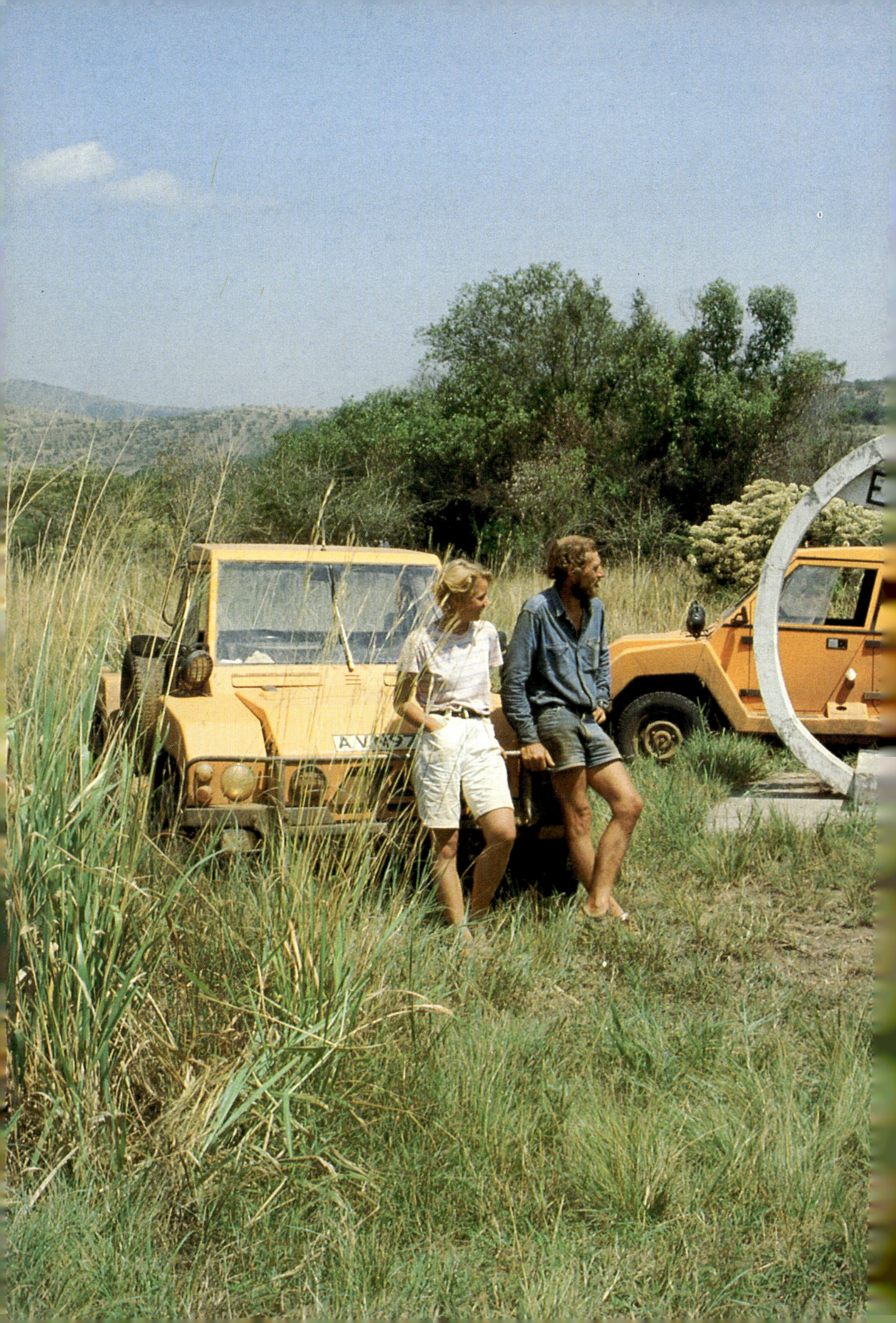

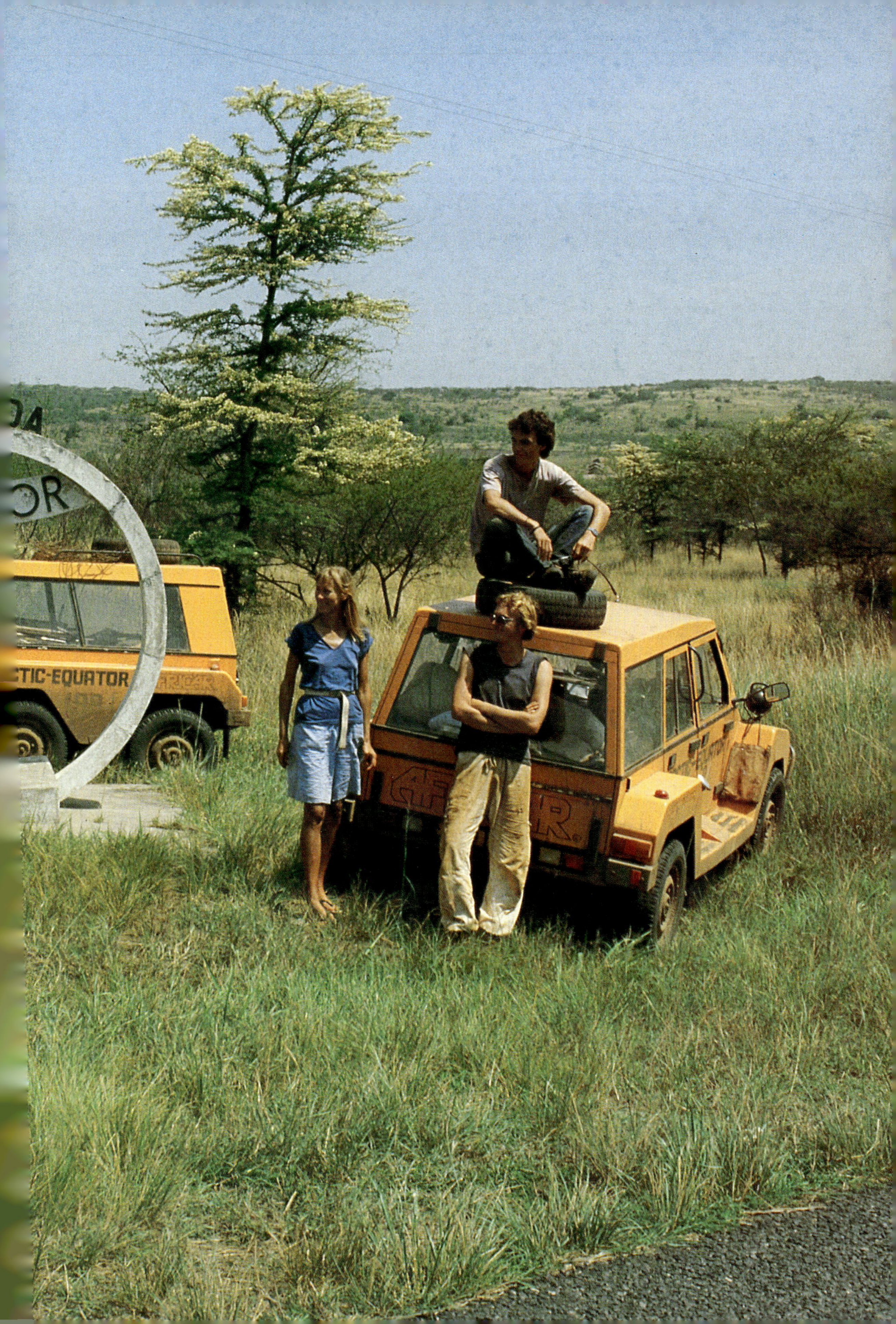
CTIC-EQUATOR

The weather clamped in. I had promised a glorious drive through the Mountains of The Moon before reaching Uganda. We saw nothing.

Well, we made it. It's really strange to be sitting here at the Equator with three wooden cars, two of which were finished the night before we left England. It's nice to be sitting here because all the prophets of doom we've met along the way said we would never make it. They said that if we got through the Sahara we'd break the cars in Zaire.

THOMAS: The main problem was Niger and Nigeria and nothing else, all this time we waited under the trees, on the road, going backwards and forwards thousands of kilometres. This gets on your nerves after a while.

CHARLES: I think the decision we took in Niamey ultimately to go through Chad,

◁ The Rift Valley in Kenya

In the Amboselli Game Park near Kilimanjaro in Kenya ▽

whatever the risks, with the benefit of hindsight, you can see it was the only decision we could have taken. You build three cars and say, right we're going to drive them from the Arctic to the Equator and prove them, you've got to get there, you've got to do it and we did what we set out to do.

LIZA: I did enjoy our time in Niger because it gave us, or gave me, an opportunity to get to know a town properly. We have been passing through the countries and towns so quickly. If I have one really memorable experience it was just standing under the waterfull at Kembe with the thunderous noise, taking a shower. For days you've been dirty and just standing under this torrent of water in the middle of the night was wonderful.

CAROLYN: I thought the desert was wonderful when I was in it. I thought the cars were doing beautifully but as the roads got worse and wetter with bigger holes, the cars seemed even better and anyway I liked that part of the trip more than even the desert.

CHARLES: I find it very hard to think this is the end of it because we're still a long way from Nairobi and I think I probably won't realize we've finished until we're sitting back at Heathrow Airport or wherever we're going.

We are still, I think, 2,000 kilometres from the coast but we're on tarmac and you know what that means. In Niger we were doing 1,000 kilometres in a day and really from here to Nairobi should be two days at the most.

CHARLES: I think the other thing is that having got this far and having done this in these vehicles, I really couldn't imagine doing another trip like this in anything else but an AFRICAR. It would be purgatory.

Relaxing near Amboselli ▷ Re–fuelling on the Nairobi/Mombasa road

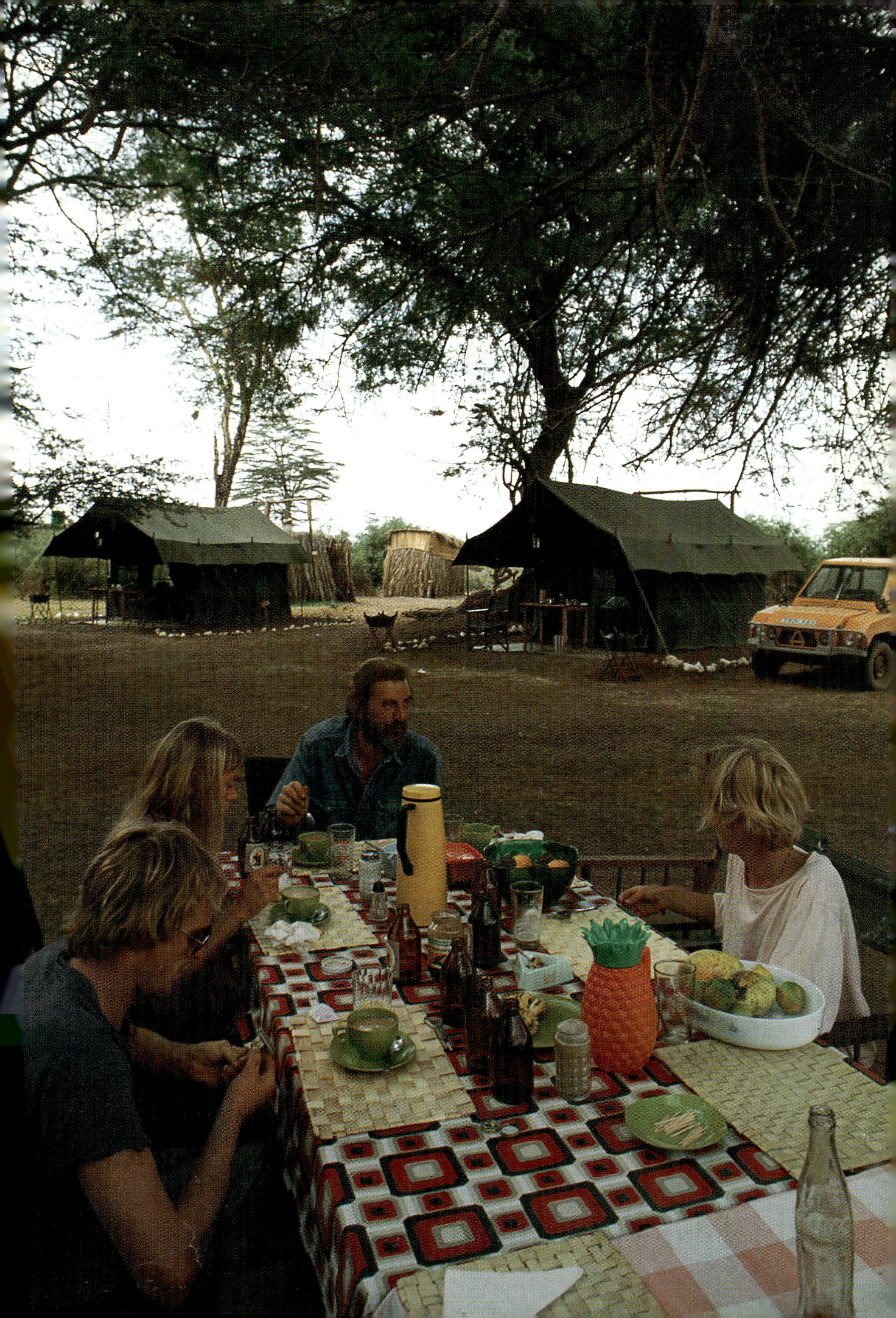

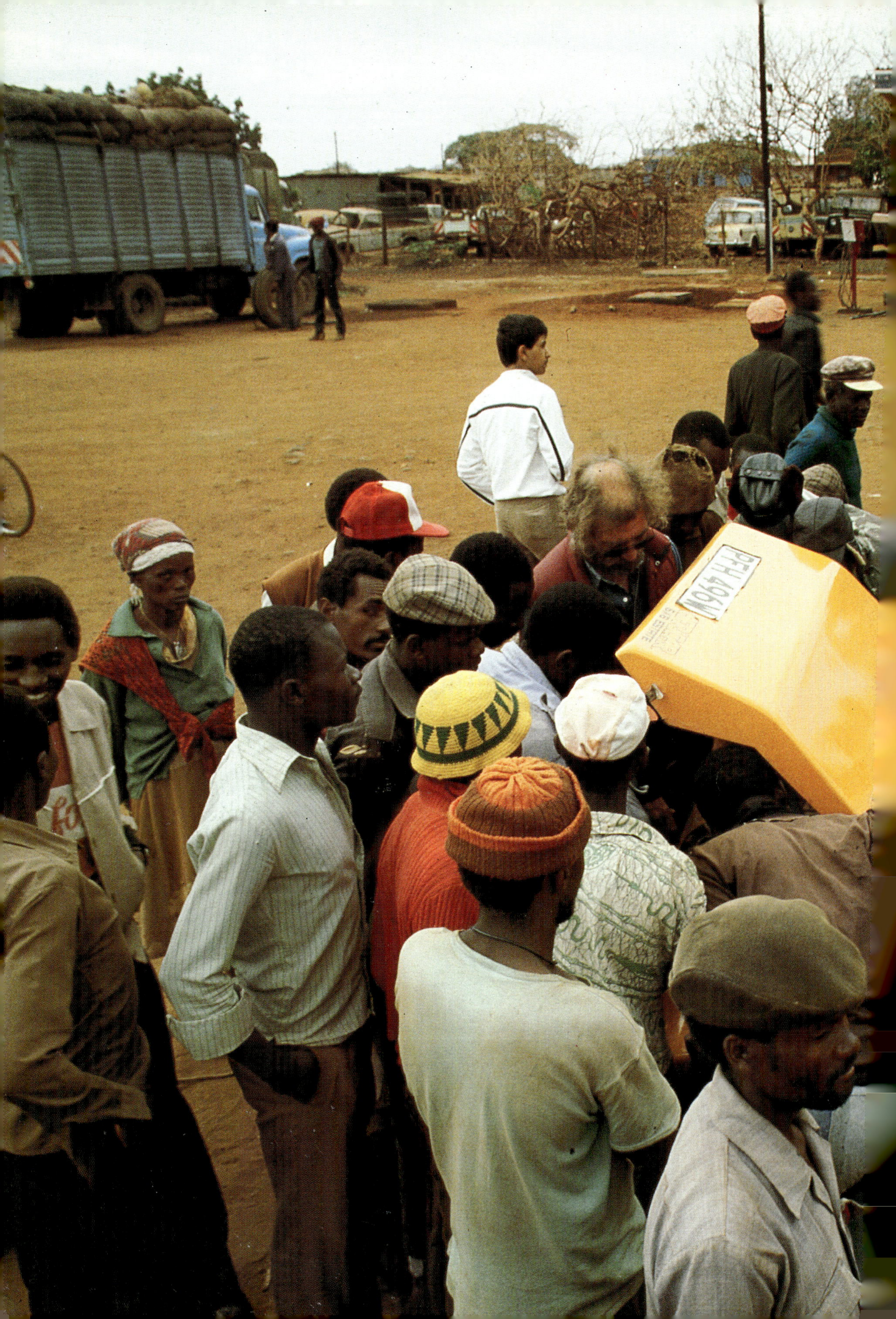

PFH 496W

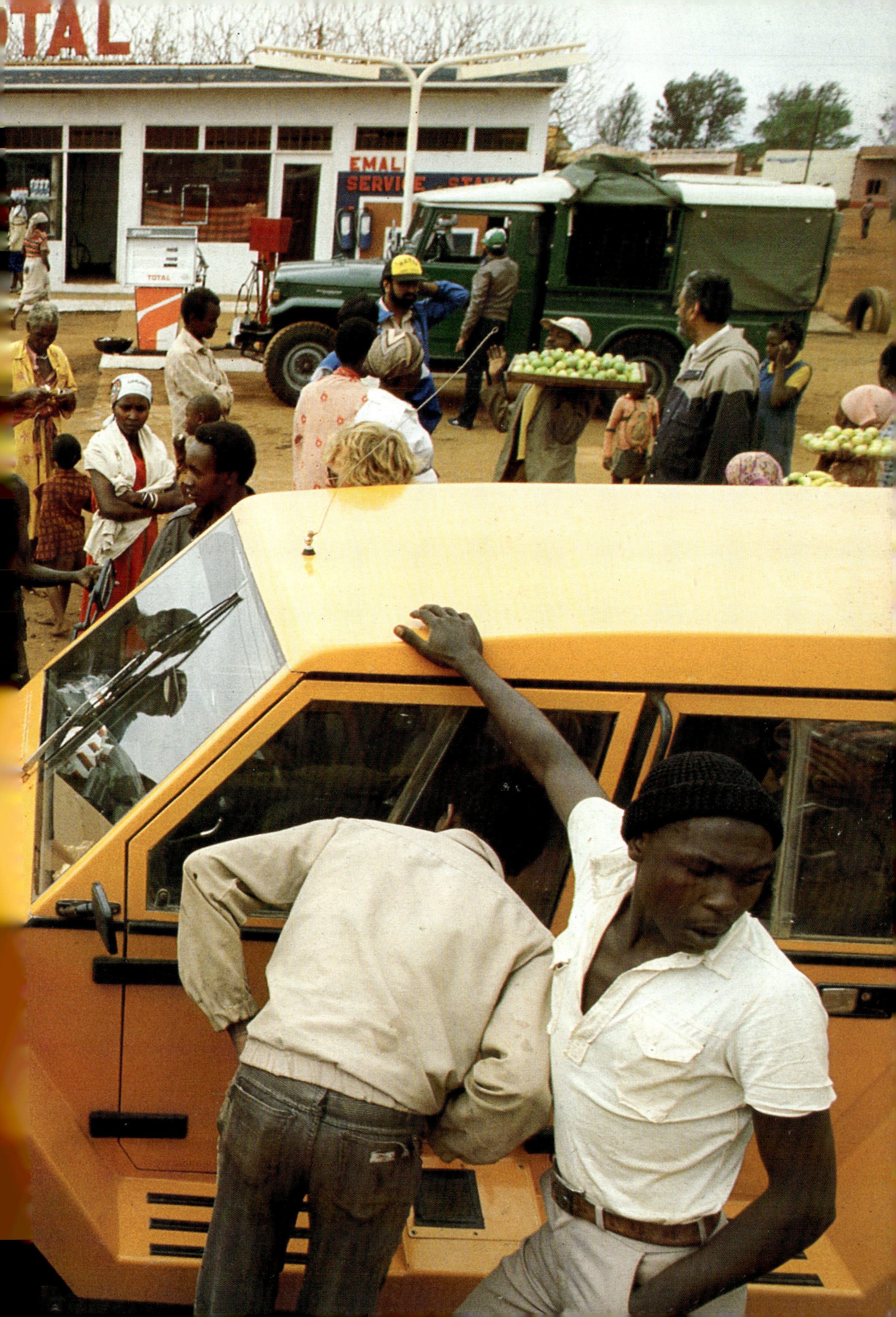
EMALI
TOTAL

I think if we're lucky we may get something of a press conference in Nairobi and the chance to get a little bit of a holiday and see some of the East African game parks, which would be nice for a change instead of feeling all the time that we have to get somewhere.

CAROLYN: *This is roll SE/70...7 and it's the 19th of July. We're at the Equator in Uganda. It's been a piece of cake.*

Let's get off and taste this nice Uganda tarmac. It'll be a very pleasant change.

Okay, start your engines, off we go.

The Uganda roads did not live up to their promise but the main routes in Kenya were almost as good as Niger. Then we were back in Europe . . . sorry, Nairobi.

The traffic was a shock and it was fraught. But not as fraught as the previous week. Arriving in Kampala the capital of Uganda we had expected to collect money wrung out of our bankers like water from a rock. Of course it wasn't there. When we had coerced our way to Nairobi it wasn't there either - it was back in Uganda.

We certainly did get our press conference in Nairobi. In fact we got much more than we bargained for.

At the press conference

The bigger the bodywork the heavier the vehicle and the more fuel it will use. We've tried to offer a system whereby somebody who normally only travels with one and a half people and carries a small amount of stuff, can have a more economical vehicle, but all within the same chassis, motor, gearbox and transmission system. That way you get a wide range of vehicles increasing the numbers and you are able to keep the price down of the basic components that make up the vehicle.

The most basic pick-up is 1,250 cc twin cylinder diesel which should give it over 50 miles to the gallon and for a farmer this is the perfect thing.

We are not, under any circumstances going to buy engines from other companies, this is the worst possible thing you can do. You then become a slave to that company and they won't let you build that engine in Kenya. They might let you assemble it but they won't let you build it.

INTERVIEWER: Where is the engine going to be built and what is the total capital investment in AFRICAR to date?

It's going to be built here. It's about three and a quarter million pounds, or something like that.

INTERVIEWER: And where's the money coming from now?

It's all come from Carolyn Hicks and myself and our friendly neighbourhood bank - and that's it.

We had arrived in Nairobi just in time for the motor show. Offered a stand, but no other facilities, we fell back on doing what, at that time, we knew best. We set up a desert camp, complete with several tons of sand.

THOMAS: We can change the body shape as we want, because the main structure is the chassis and what is put on top doesn't matter at all. You can bring it out to the full width, can bring it up in height, you can make it into a "matatu" for 17 to 20 people.

SPECTATOR: And what is the chassis? Is it a fairly conventional chassis then?

THOMAS: No, it's wood. We have metal beams at the front and back on which the suspension arms are mounted. This is attached to the plywood but the whole chassis itself is made of wood.

And now we go back to the drawing board and we do the next series of cars with all the lessons that we've learnt. At the same time our own engine and gearbox are coming along. But it's a very big project, it's not a kit car, it's not something that's put together from other vehicle's components.

It is very hard to do because people are sceptical, but that doesn't really worry or interest me, because I'm doing it anyway.

INTERVIEWER: What made you decide to build the AFRICAR

I decided I didn't want an ordinary car and didn't want something like a Land Rover, a big heavy four-wheel drive. What I wanted was

◁ **Our end of trip conference at the Norfolk Hotel, Nairobi**

△
A prize for AFRICAR in the Concours d'Elegance at the Nairobi Motor Show

After snow and ice, European motorways, desert and jungle tracks, the last test surface was East African Savannah
▽

something in-between designed for Africa. In other words an AFRICAR.

Invited to enter the concours d'elegance, I fear that we insulted the judges. But we all felt the prototypes were looking pretty good just three days after the end of the trip.

JUDGE: No bloody effort's been made to clean it!.

ANNOUNCER: Another crowd puller, Carolyn Hicks and Tony Howarth's AFRICAR. All the way from the Arctic Circle. Well done Tony, (presenting second prize in the Pick-up Class).

I had first conceived the car for Africa concept while travelling in Zambia and Tanzania 20 years ago. Subsequently many of the ideas developed from experiences throughout the continent, but especially in Kenya.

Driving in the Masai Mara below Kilimanjaro the AFRICARS had come home.

We've found one of the world's perfect places. it's called Chula Camp. It has a river, shade, all those things we looked for in the desert. Here we can relax and try and recuperate from what actually has been a very hard trip.

Africa always produces pressures, whether it's borders, distances without fuel, roads that are almost impassable or rains when you least expect them. This is what has made this test so good.

What we've really got to do this afternoon is to find an elephant for Carolyn.

CAROLYN: I don't believe they exist.

Well, either this afternoon or tomorrow because last time we were here I spent two weeks trying to find her an elephant.

CAROLYN: We chased throught the Aberdare Mountains, the famous elephant walks, every place - nothing.

After a two year fight for survival against indifference, contempt for the Third World and substantial vested interests, AFRICAR is now controlled by a non-profit foundation. 'Worldwide Appropriate Transportation Trust'.

8
AFRICAR 1987

On our arrival in Nairobi we considered the Arctic to Equator journey had been a success. To have undertaken the same journey with four production vehicles would have been courting disaster – undertaking it with prototypes was rash, to say the least. The vehicles had shown they could cope with the conditions and that their fundamental structure was more than strong enough. In particular, the suspension, which was the element we were really testing, had worked superbly. We had also found the AFRICARS to be ergonomically satisfactory, comfortable to drive and effective to live in. They were capable of carrying the varied loads we needed within their structure without additional roof racks. Only extra spare wheels had been bolted to the roofs of the wagon and six–wheeler.

Tyres were, in a way, our greatest success story. During the entire journey we only had slight sidewall damage to one tyre. The suspension and the soft wheel rims had protected us at all times. In fact, the six–wheeler's rear tyres were only half worn by the time we reached Nairobi and the tyres of the other vehicles were about two–thirds worn, suggesting a possible tyre life of 50,000 kilometres under harsh conditions.

On the journey Thomas and I had gone a long way to establish the necessary parameters for a series of AFRICAR engines and for a unique four–wheel drive system. In the few weeks we spent in Kenya, the new 'AFRICAR gearbox' was drawn up in reasonable detail.

Although we knew that funding was in a critical state, given the success of the journey and the existence of the basic design for the final AFRICAR elements, we expected to find backing to take the project forward, once we returned to England. Also, by this time, over thirty countries around the world were expressing serious interest in the possibility of manufacturing the vehicles. Many individuals were offering to buy an AFRICAR as soon as we could have them available.

In the past twenty years the industrial climate of England has been bleak. The emphasis, orchestrated by so–called financial institutions, has been on the service industries and on pure financial manipulation. An attitude of: 'Why bother to go to the trouble of making something when you can turn over a bigger profit just by acting as middleman?' A short sighted attitude that has all but destroyed the entire manufacturing base of the British Isles. It is also a curious attitude as it is evidently based on no 'market research'. The only countries which exist effectively on service industries are poor Third World countries, or one or two tax havens. All the other developed nations have substantial and progressive manufacturing sectors. It is no coincidence that Sweden has just about the highest gross domestic product in the world. It actually is, on a per capita basis, by far the world's largest vehicle exporting country. In Sweden 70 percent of vehicle output is exported, unlike Japan which exports only about 50 percent.

This attitude towards financing industry has resulted in very few actual manufacturing business start–ups in the United Kingdom between the mid–1960s to mid– 1980s.

So, on our return, we were disappointed, but perhaps not surprised, to find that no–one known to us was either impressed or interested in our achievement. We found ourselves, quite literally penniless, virtually unable to survive on a day to day basis. We could see no value being put on the work we had done and absolutely no future for the concept.

At the end of September 1984, Theo Agnew, a young Englishman working in Australia, walked into our office and offered to buy a prototype. Under normal circumstances we wouldn't have considered doing this, especially as the cars had just completed such a gruelling journey. We couldn't even be sure of their safety. He wanted Doug Stewart, a veteran Australian endurance rally driver, to compete in the Himalayan Rally and then to have the car shipped out to Australia. He had organised free transport to India for the vehicle and for myself and Thomas as mechanics.

Quickly we managed to raise another few thousand pounds of sponsorship and, on the basis that the vehicle would be prepared for the Rally and therefore thoroughly checked over after its African experience, we agreed.

I decided to go with the vehicle to India to co–drive and then to continue to Australia,

The headquarters of AFRICAR. Built as a global village amidst reminders of the environments in which AFRICARS must function

New Zealand and the Southern Pacific area, because it was there that we had the greatest interest in AFRICAR at that time. It seemed to be our only hope and only possible course of action.

With two weeks to prepare the car for the Rally, it was done with frantic haste. A Volkswagen fuel–injected 1700cc engine was substituted for the Citroen unit to give us more power. A year earlier we had competed in the same rally with a Citroen engine and it had run well in the first few stages before falling out with fuel pump problems. The standard AFRICAR pick–up had shown that it could make the fifth fastest time and had run in around seventh position, so this time we thought a little more power might lead to a better result.

Rallying may seem a long way from the essence of the AFRICAR concept. But I had learnt that people living in the Third World naturally wish to be as proud of their vehicle as people in industrial countries, particularly as it has generally cost them an exorbitant amount of money. Credibility and prestige at all levels is as important to the future of AFRICAR as it was to Nissan or Audi. It would be insulting to the customer if an AFRICAR vehicle could not prove itself competitive in serious endurance rallies, in the way Peugeot had done in the 1960s and 1970s with their very ordinary models. Their rallying success did not turn the product into toys but reinforced their stamina and capabilities.

On the starting line in Delhi we had a slipping clutch due to hasty preparation. And worse, we had found that turning the lights on at night substantially reduced engine power. One of those gremlins that creep into a vehicle from time to time and to which we never found a logical solution.

With Doug at the wheel, we set off on what turned out to be a fraught but entertaining event. On the first competitive section we were about 500 metres from the end, and set to make equal fastest time, when a driveshaft joint seized. From that point on we gradually worked our way through the four ex–African trip driveshafts and slipped back to the tail end of the field. By the third day we had replaced all the driveshafts, the car was running well and with some spirited driving from Doug we managed to make up nineteen places on three sections.

We were at the mid–point of the Rally when Mrs Gandhi was assassinated and the whole of Northern India was plunged into turmoil. The Rally was stopped and cancelled so our hopes of moving into the first ten positions were never realised, but we did collect the prize money for being second in our class.

The journey back to Delhi under army escort was adventurous and appalling. Burnt out buses and trucks littered the roads, we were threatened with guns, and the whole time there was a feeling that the entire nation was on the point of boiling over into anarchy. I abandoned my planned visit to a number of Indian industrialists and vehicle manufacturers and immediately flew on to Australia.

The next four months were a period probably best forgotten in the story of AFRICAR. Without resources, I found myself in an area which I had not previously experienced. There was enormous interest in AFRICAR but I did not have the strength to carry it through to any logical conclusion. I also found myself up against the negative influences of local vehicle distributors. I learned an important lesson in that although major car manufacturers may regard these markets as peripheral and possibly not cost effective, the local distributors of Japanese, European and American vehicles are, of course, committed to importing and retailing their product. The idea of an AFRICAR factory making vehicles locally, and saving the nation foreign exchange, is an anathema to them and they are likely to use all influence possible to stop such a nationally advantageous development.

After the rally much of my time was spent in Fiji and I only wish it could have been under better circumstances. In the end I could see no future in pursuing AFRICAR there or in New Zealand, although there had been some good signs coming out of Australia. By now it had become completely impossible to continue and so I used my sad collection of dubious low–priced tickets to make a weary five–day journey back to England. Right then, if I had had a tail it would certainly have been between my legs.

In England I found that Carolyn had somehow, once again, managed to negotiate a kind of truce on the financial position. Although uncertain, it looked as if we could just possibly find a way to continue. The Spring, Summer and Autumn of 1985 are memorable for two things. Twelve hours a day sitting at a desk working on what was seen then as the ultimate AFRICAR Feasibility Study and an acute lack of funds, to the point of not even being able to eat properly.

Thomas Marx, who had joined us in the Sahara as mechanic, stayed with us while he and I produced a 600 page document that went into fine detail on design, production plant, vehicle costing, capital costs and a complete start–up scenario for both developing and developed countries. It was

Thomas Marx's drawing of my concept for an opposed–piston, supercharged, two–stroke AFRICAR engine ▷

My initial drawings of the full production AFRICAR in 1987
▽

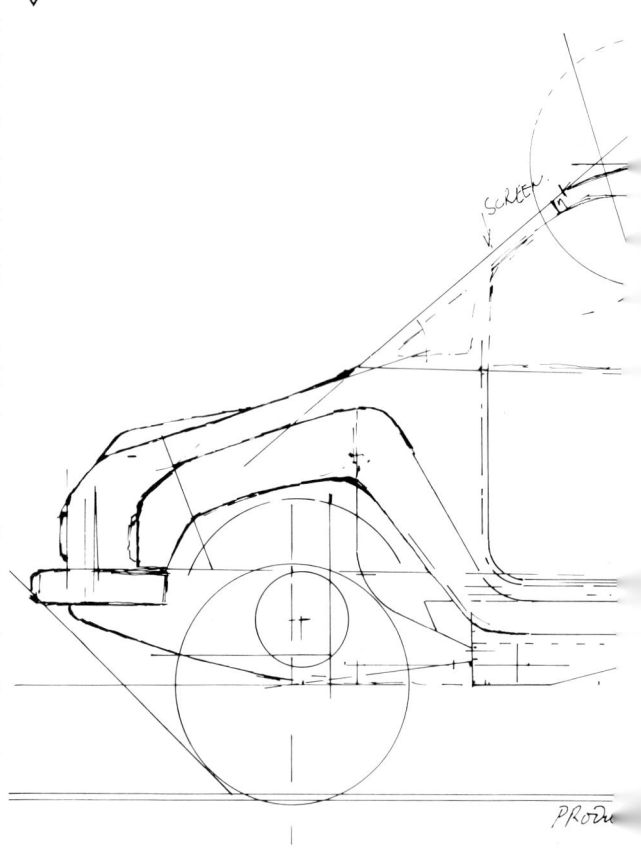

an element of AFRICAR that had been done rather inadequately before we left on the Arctic to Africa journey but now had become essential if we were to get anyone to take us seriously. We finished this massive task, done without computers or memory typewriters, at the end of November. Thomas returned to a job in Germany.

Armed with carefully researched facts and figures I was able to produce a potential price list for production AFRICARS and, with the help of Dick Vine, created a catalogue which has since then become the basis for AFRICAR marketing. It is amazing what can be done with a typewriter, a pen, ink, paper and a Xerox machine.

One evening in March 1986, Dick came to see me with a friend of his, Mike Williams. They had asked me, rather formally, for an appointment. In effect they sat me down and said: 'This won't do and we are going

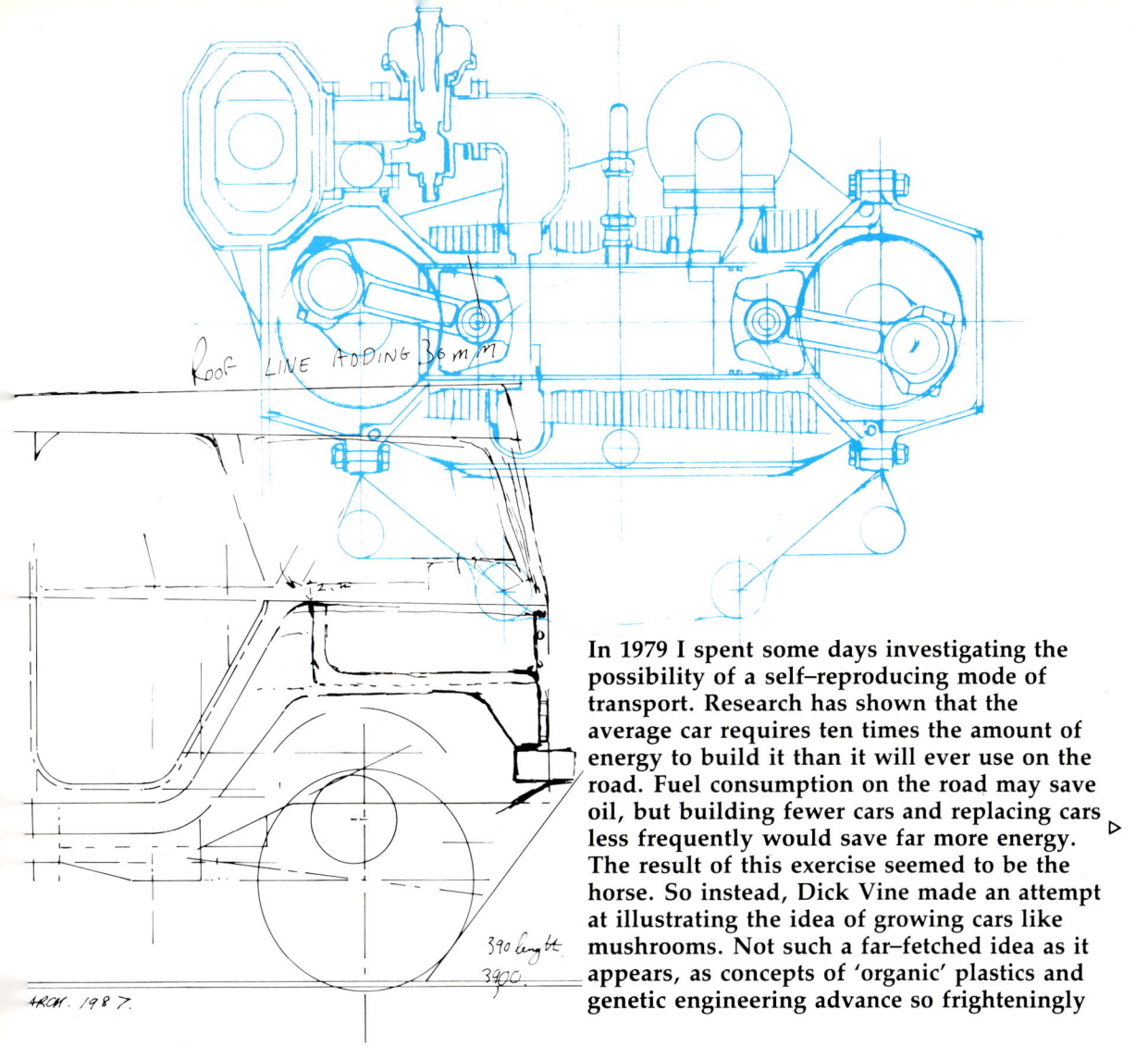

In 1979 I spent some days investigating the possibility of a self–reproducing mode of transport. Research has shown that the average car requires ten times the amount of energy to build it than it will ever use on the road. Fuel consumption on the road may save oil, but building fewer cars and replacing cars less frequently would save far more energy. The result of this exercise seemed to be the horse. So instead, Dick Vine made an attempt at illustrating the idea of growing cars like mushrooms. Not such a far–fetched idea as it appears, as concepts of 'organic' plastics and genetic engineering advance so frighteningly

to do something about it. How much do you need to get you into some sort of factory, have some presence, and get this project moving forwards once and for all?'

It was such an unlikely question for Dick to ask, I had never perceived him in the role of entrepreneur. Before I could answer they said: 'Look, we can come up with about 20,000, will that be any good?' It took some minutes for that to sink in and then I realised that if I added to it the 10,000 I expected to inherit from my mother's estate, it would be a total of about 30,000. This was, after all, more than had been spent to build the first prototype. On the other hand, as a sum of money in relation to the hundreds of thousands that had been spent between the first prototype and the Arctic to Equator cars, it seemed pitifully small.

I think it was their enthusiasm and determination that turned me around. 'Why not?'

I said, 'I'll go out and find a rent–free factory and we'll get it going.' It was that simple. Just two and a half years after our line of credit had been cut off at the end of 1983, AFRICAR seemed to have a future.

I had previously opened dialogue with the Scottish Development Agency which wasn't getting anywhere, so I had decided to look at some of the development areas in England and chose, for geographical reasons, Warrington, near Manchester. But, after my mother's funeral I had gone to look at my old home town, Lancaster, and had taken the time for a talk with the Development Officer for the City.

A call from the Lancaster City Council came a few days after that crucial evening meeting and I was asked to discuss the project with the Town Clerk. I think that without the backing from Dick and Mike, I would never have been able to put on such

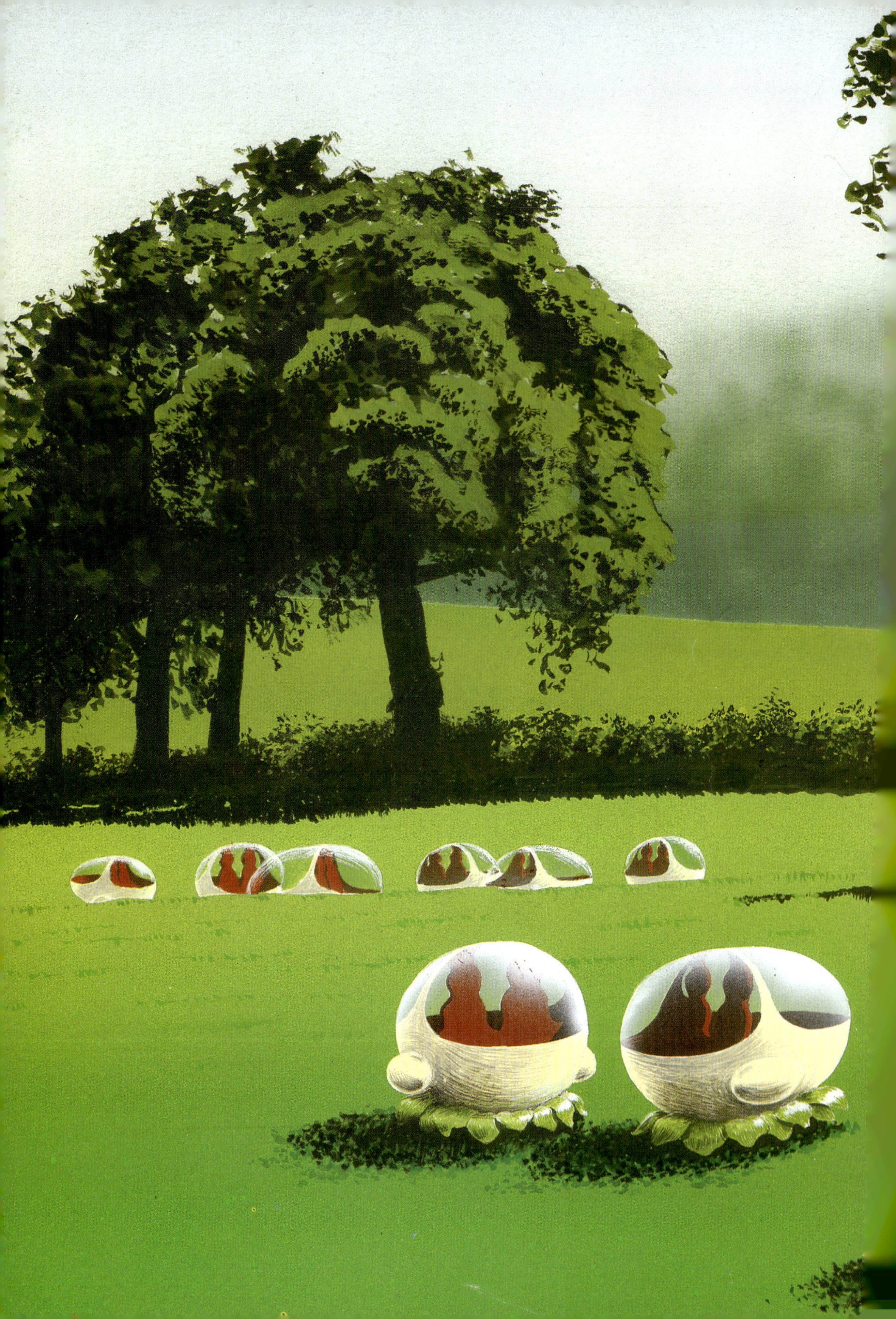

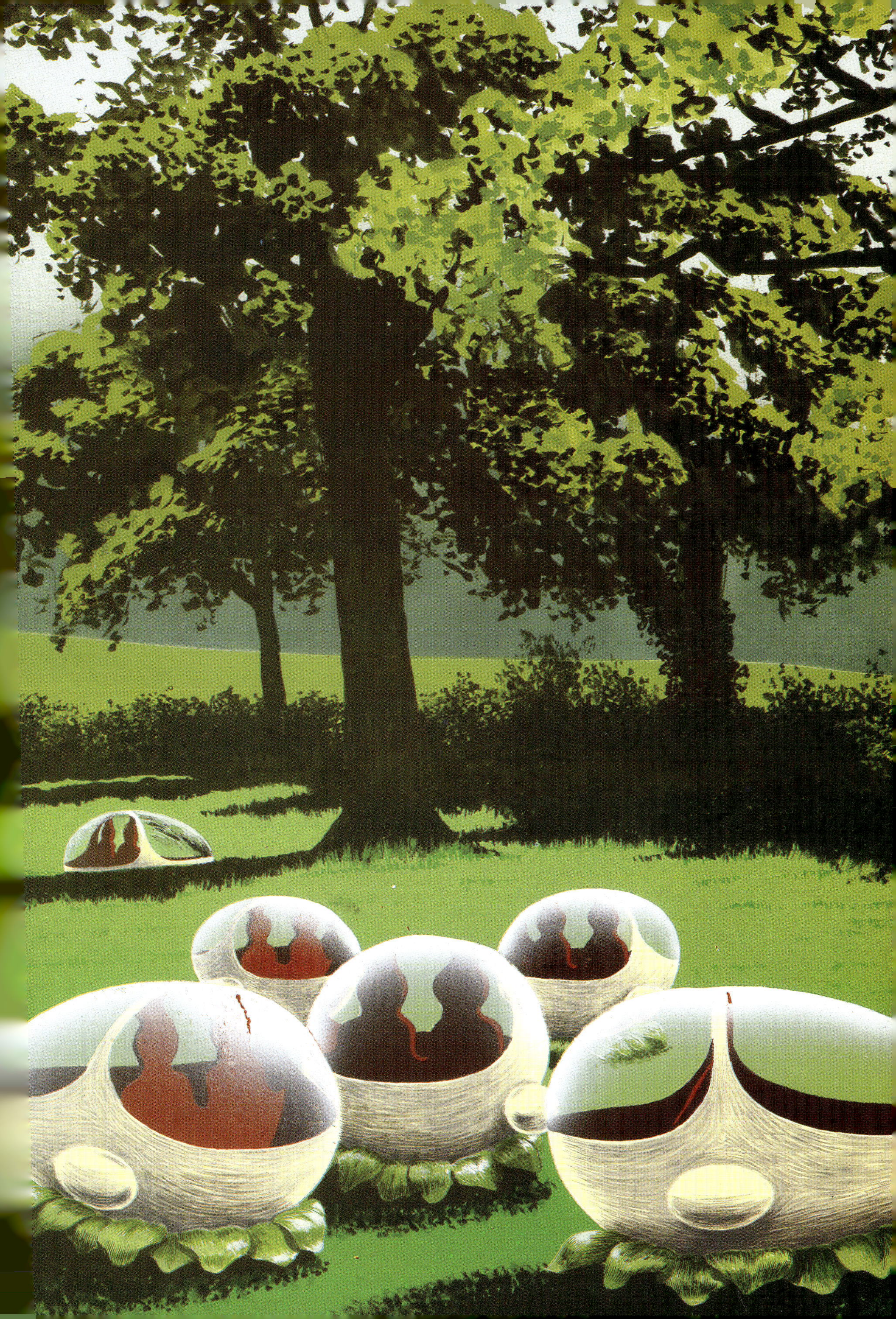

a brave face at that discussion, although another very positive development was coming about as a result of the catalogue. Already several vehicles had been sold against a possible production date in 1987. I was also getting greater interest from Australia, plus the all important fact that the first contract for AFRICAR distribution had been signed, for Bangladesh, Burmah, Bhutan and Nepal.

Lancaster, in the North West of England, is not considered a development area by the Government. However, its unemployment rate and associated problems are as bad as anywhere else in the country. Never an industrial heartland town, it still had a basis of industry including artificial fibres, linoleum and other adhesive plastic sheeting, chemicals and furniture manufacture. Most of this activity ceased in the 1960s and 1970s. It was here, in the workshops of Waring & Gillow, a fine furniture manufacturing business, that I remember seeing plywood aeroplanes being glued together during World War II. Parts of the Mosquito Nightfighter and the Horsa Gliders were made in Lancaster at a time when my father worked for the company.

With no special development or hardship designation, the City Council had little to offer, but, and this was a very big but, what little they had, they offered. At the end of that meeting I made an agreement that AFRICAR'S U.K. base would be in Lancaster on a permanent basis. In return I had the offer of adequate factory space to get started, additional space for expansion and potential sites for building on at a later date. There was also the promise of assistance with finding a suitable workforce, a cash grant to cover re–location costs and a general assurance that every assistance would be given to me in this start–up situation. There was a 'can do' attitude in Lancaster which I hadn't come across anywhere else in this country.

It made me wonder if my own 'can do' attitude to AFRICAR and other things has something to do with the fact I come from the Lancaster area.

From that day onwards AFRICAR started to accelerate. Distribution contracts were signed up and paid for, with sometimes as many as six countries within a month. An Australian agreement for a full manufacturing plant was signed in October 1986. And small administrative headquarters were established where work went ahead with productionising the prototype vehicles which had given us such good service since they were completed in 1983.

An official and formal structure for the AFRICAR organisation emerged which was influenced by the experience of the past few years. I decided, with Carolyn, to place the control of the AFRICAR concept with a non–profit making foundation, thus ensuring that the long–term objectives of multiple manufacturing centres, especially in countries that have no manufacturing tradition, would not be lost because of financial constraints or demands from future investors. At the same time it was decided that the local operations, one of which would be in Lancaster, another in Australia and others under discussion in various parts of Africa, Central and Southern America and Asia, would be open to support from local investment and run for profit for the local community. They could also be backed by grants and soft loans from major international financial institutions such as the World Bank or the International Finance Corporation. In addition, by establishing the Foundation, there was the possibility of actual public subscription to support the development of AFRICAR in the places that most needed it.

To suggest that once the negotiations with the Lancaster City Council were over, the rest was history, would be inappropriate. That was just the beginning, for me seven years down the line, but still the beginning. The one thing that I was certain of was that AFRICAR was no longer just a concept and a development dependent entirely on two people. Its future would really depend on the public response, wherever that might be.

AFRICAR was always seen as a solution to some of the problems of Third World transportation and off–road travel throughout the world. It was certainly never seen as the only solution. With its technology developed, it can now be offered to the world and the world has a choice as to whether it accepts or rejects it.

I need a holiday!

APPENDIX

If ever the wheel has to defend itself, the bicycle is its justification. Throughout the world the bicycle has become the car's closest rival in putting people on wheels. At the last count there were believed to be some one hundred million bicycles in use. Its advocates claim that it is the most efficient form of transportation developed to date, for the bicycle offers greater sustained speed and far greater range for the same load–carrying capacity than the horse. Yet a very small expenditure of energy is required to propel it, perhaps even less than riding a horse.

The bicycle has only been around as long as the car, which is surprising considering most of its basic technology, including gears, has been known for thousands of years. The simplest 'hobby–horse' came as late as the mid–1700s and when the Frenchman, the Count de Sevrac, produced an improved version in 1791, it was still without pedals or steering. In 1816 the German Baron Karl von Drais added the steering. In 1839 Kirkpatrick Macmillan produced a machine with pedals, powered directly to the front wheel, without drive chain or gearing. These 'refinements' were incorporated in the velocipedes which were commercially produced from 1883 onwards by Pierre and Ernest Michaud, claimed in a centennial French postage stamp as the 'creators of the modern bicycle'.

The invention of the pneumatic tyre by R.W. Thomson of Edinburgh in 1845 and its exploitation by John B. Dunlop of Belfast from 1880 onwards, was initially crucial for the bicycle. Developments in metallurgy and in manufacturing precision ball–bearings were of great importance in the metamorphosis of the bicycle from the un-comfortable toy of the rich to the practical means of locomotion for the masses.

In 1979 the bicycle joined the space age and took to the air. Charles Allen's astounding act (not forgetting the achievements of the designers and builders) in pedalling the Gossamer Albatross twenty–two miles across the English Channel is surely, in terms of ecological and technical achievements, one of the greatest successes of the century.

But pedal power is only pedal power. It was an independent source of motion that we sought, the means to indolence in our lives.

Evidence that experiments to provide motion with heat, magnetism and electricity long preceded our own industrial revolution can be found in the London Science Museum. The museum staff have constructed a working model of a simple steam turbine based on a device called an aeolipile, designed by the inventor, Hero of Alexandria in the first century AD.

There are earlier references to land–yachts from Egypt and China, probably the first practical self–propelled vehicles. But it is not until the 16th and 17th centuries that there are any detailed records of self–propelled land–locomotion in Europe.

The beaches of the Netherlands provided the testing ground for a number of land yachts and there are detailed records of the achievements claimed for machines designed and built by the Flemish engineer, Simon Stevin. A contemporary chronicler claimed that in 1600 one of Stevin's vehicles with twenty-eight people on board travelled 67 kilometres in two hours – a speed of about 20 mph.

The land yacht with its sail motor did not die out as some commentators suggest; it became the ultra–efficient sand yacht of today. The modern sand yacht is capable of speeds of up to 70 mph, is not limited by the direction of the wind and is used for highly competitive circuit racing.

More exotic attempts were made to harness the wind, such as that by Englishman George Pocock in 1882 who devised his Char–volant, a carriage pulled by a series of steerable kites (an idea that may come back into vogue for some specialised uses in the 1980s). The vehicle, patented in 1826, seemed to have worked extremely well, achieving speeds of between fifteen and twenty miles per hour and even putting the stage coach to shame on the Bath to London road. Pocock wrote in his diary:

'Just as we were preparing to depart, the London stage came by. It was almost fifteen minutes ahead when the Char–volant set out, but after a run of four miles, the Char–volant was alongside it. Its efforts were fruitless and after a run of ten miles, we entered Marlborough twenty–five minutes ahead of the stage coach.'

No mean achievement in navigation to anyone who knows that road with its many corners and trees.

Towards the end of the 15th century, Leonardo da Vinci was designing all kinds of elaborate motor–driven carriages and war machines. For want of a usable motor, Leonardo used springs in tension to provide motive force. He even recognised, and solved in a primitive manner, the problem of the differential. That is the means whereby the outside wheel can travel further and faster than its inside counterpart when a driven axle goes around a corner. Some cars did not compensate for this problem right up until the 1940s.

From about the end of the 15th century the search was on for a motor. German and Austrian clockwork artists made full-size clockwork carriages that were reputed to travel nearly a mile on a single winding. In 1740 Jacques de Vauconson's clockwork carriage ran on the streets of Paris.

In 1875, when the petrol driven car was already on the road, a serious demonstration of a clockwork tram, wound up by a stationary steam engine, took place in Southwark, South London. The principle is not dissimilar to the use of a flywheel as a motive force, a method used today in a number of public transport networks. Generally on buses and trains, which have steady or frequent access to a regular basic power source, the flywheels, accelerating to thousands of revolutions a minute, propel a vehicle through sectors where its regular electric power is not available. The system has also been utilised as a means of harnessing the energy which would have been lost through braking.

Clockwork cars have, of course, survived as toys. More and more toy cars however, are being powered by electric motors, have rechargeable batteries and are controlled by electronics. It is curious to think that the cars of the future may already be in the hands of our children. The historical precedent for this is well established. The first steam–powered vehicles were made in the 17th century as models and toys.

As early as 1629 the Italian Giovanni Branca designed a steam engine. A few years later Otto von Guericke invented an air pump comprised of metal pistons, cylinders and connecting rods, the basic components of the reciprocating engine. This same genius made rough sketches in 1661 of an internal combustion engine.

During the same century Sir Isaac Newton invented a simple device to be propelled by a jet of steam and prophesied that one day man would be able to travel at fifty miles per hour in steam–powered carriages. An important advance was the atmospheric engine designed by the Marquis of Worcester. As early as 1678 a Belgium priest in China, Ferdinand Verbiest, made a steam carriage suggestive of the modern turbine. A model based on his plan exists in the National Museum of Science and Technology in Milan.

This was the same year that Robert Hooke patented a steam engine in Britain. Following the invention of a steam pump in 1682, Denis Fapin of France built a model of such a machine and in 1698 Thomas Savery obtained a British patent for one.

These men and their successors built upon their achievements and failures for a further century before steam was first used for practical locomotion.

Refinements such as the multi–tubular boiler, rack transmission, the connecting rod and crankshaft, multi–cylinders, change speed gears and the spring-mounted engine brought further advances. All these indicate that here, as in so many fields, it is impossible to credit any one person, group or nation with the crucial element of an invention.

Nevertheless, a name and date are justly recorded in the history of transportation. It was in 1769 that the first successful steam–powered vehicle made a short journey. The French engineer, Nicolas Cugnot, demonstrated his tractor to the French army and created enough of an impression for them to commission a second and larger vehicle. In 1770 the twin–cylinder Cugnot also ran under its own steam but proved so difficult to control that it knocked down a wall and in doing so became famous for being involved in the first motor accident.

David Burgess Wise, in his book The Motor Car, points to the coincidence that this first French motor vehicle, with its front wheel drive and engine mounted forward of the front axle, gave us a foretaste of the almost universal modern layout, as propagated for the ordinary motor car largely by the French company Citroen.

Before the end of the 18th century steam engines had been successfully developed in Britain, France and the United States for land and water borne transportation. Steam buses were already running on the streets of Paris and in 1804 Richard Trevithick of Cornwall had demonstrated the practicality of steam for a locomotive on rails.

There followed in the early 19th century a flood of steam carriages of all shapes and sizes. For a while steam locomotion on the roads seemed to be establishing itself as the transportation system of the future. By the early 1830s thousands of fare paying passengers had used this means of transport, while only a handfull had paid for journeys by steam engines on rails.

What is claimed to be the first long distance motor journey by road was undertaken in 1829 by Sir John Goldsworthy Gurney's 18–seater, six wheeled steam coach, which travelled for eighty miles. Two years later he inaugurated the first inter-city service by steam bus – from Gloucester to Cheltenham. Also in that year the New York and Harlem tramway first used a cable system powered by steam.

During the same decade came the first mainly steam–powered Atlantic crossing and the first propeller driven steamship. In 1852 steam power was applied to the navigation of an airship. Taking account of the efficiency of human and animal legs as a means of converting energy into locomotion, David Gordon and Sir Goldsworthy each invented steam vehicles based on the use of backward–thrusting mechanical legs. (Similar vehicles have been designed more recently to aid man's movement as he explores the Moon and other planets.)

If the conventional steam vehicle appeared to offer the greatest hope for road transportation, why is it that most inventions in this field did not even go into service and those that did rarely survived for long? The first fifty years of the steam carriage saw high hopes continually being dashed into ignominious defeat as the gleamingly ornate new carriages were towed back to base by a team of horses.

Although few of the steam entrepreneurs of the time would have admitted it, the compromises of the transportation equation caught up and overtook the technology that was available.

The steam engine offered a motor, curiously and ironically, a really good motor for driving a horseless carriage. Like the electric motor, the steam engine requires no clutch or fluid coupling to the wheels, it is directly connected. When the wheels stop, the engine stops and sits there quietly waiting until you tell it to go again. Unlike the internal combustion engine the steam engine does not have to rev to produce the all important torque, the turning or twisting effect that is needed to turn the wheels and accelerate the vehicle. A double–cylinder, double–acting steam engine has the same number of power strokes as a four–cylinder petrol engine, to every revolution. It was too good to be true.

The problem was not in the engine itself but in the supplies it needed; the boiler, the water, the fire and fuel for the fire. If the vehicle was to have any range at all these items became very heavy. Such weight needed a massive chassis and strong wheels to support it. Add to this the weight of the passengers and the luggage and you had a monster. Try and cut down weight and you reduced range, safety and reliability. Then there was the time taken to get up steam and the danger of the boiler exploding, which happened. Later these problems were eliminated at the cost of enormous complexity with the invention of the flash steam generator.

These clumsy vehicles drove themselves through their wheels. Few vehicles today could manage the roads of the early 19th century, wet or dry. The steam carriages certainly could not, having neither the traction nor braking of a team of horses.

Steam locomotion eventually found its home on the railways, in heavy traction engines for farm and industry, and at sea. The technical advances in steam locomotion and permanent way construction, then the consequent economic power and finally the resulting political influence, all gave the railways the lead in land transportation, which they were to retain for almost a century.

The smooth low–friction railroad track made the weight of the steam engine an advantage while the tremendous potential of smooth torque led to long, highly efficient trains.

Isaac Newton's prediction of fifty miles per hour was exceeded on rails before 1850 and the steam trains stayed widely in service until the 1950s, when they became unviable, partly because of their poor thermal efficiency (fuel consumption) and partly because of the world's addiction to oil.

Steam offered self–propulsion but it demanded compromises on every other count – cost, comfort, weight, size, safety, reliability and range. For one and a half centuries the promoters of steam tackled every one of these problems and with great ingenuity beat most of them. But even the lightweight steam cars that were produced right up to the depression could only offer silence and smoothness at the price of acutely limiting range and poor fuel consumption. As a result, steam failed on the roads.

I can't leave the subject of steam without mentioning one probability and one certainty. The reciprocating steam engine is still a very good way, especially from the driver's point of view, of powering a motor car. The promoters of steam have not lost faith and are now pushing another huge advantage of the steam driven vehicle; it can use virtually any fuel that will burn from wood to coal to waste paper, old candle ends and string. It will thrive on any type of oil, vegetable, mineral, thick or thin. We may very well see steam cars on the road again.

On a different level we are seeing and will see more of a certain type of steam car. The electric car takes its power from the mains supply, the batteries are only for storage. Most power stations use steam turbines to turn the electric generators, so in a sense, the electric car is a steam car.

The electric car is not a new idea, it was in fact, the next hope after steam and was produced in parallel with the petrol engined car for twenty years and more. As early as 1837, Robert Davidson of Aberdeen had built an electric carriage that embodied all the elements of the modern electric vehicle except a rechargeable battery. The discovery of the lead–acid accumulator made the electric vehicle practical. In 1881 the Paris Omnibus Company introduced the first commercial battery operated vehicles, and in London, the first electric self–propelled cabs were run between 1897 and 1899.

In 1889 the first world's land speed record was set at 39 mph for the flying kilometre and a year later the figure was raised to 66 mph by the Belgian, Camille Jenatzy, driving an electric streamliner. The electric car speed record now stands at 175 mph set by Roger Hedlund, driving his Battery Box at Utah Salt Flats in 1974.

Right from the start the much vaunted virtues of the electric car were apparent. As James Paterson noted in 1905 in The History and Development of Road Transport, 'Electric cars are clean, silent, easy to drive, very reliable; in fact, more civilized than any other self–propelled vehicle.'

The compromises, then as now, were the cost of the vehicle combined with the cost of battery replacement and range. Initially these compromises only existed relative to horse–drawn vehicles. The electric car could compete with other mechanical conveyances.

In the United States at the dawn of the 20th century, steam and electric cars were competing for leadership, each with about forty percent of the market, while the internal combustion engined vehicles trailed in third place, with about twenty percent.

Yet by 1908 petrol–driven cars had achieved ninety-five percent of sales, while electrics had three percent and steams two percent. Perhaps even more significant: the 1900 figures were based upon total automobile sales of just over 4,000. By 1909 the total was over 126,000. The age of the automobile from that time until today has meant the petrol–fueled, internal combustion engine.

In order to understand what happened to the electric car, the steam car and the horse drawn carriage and to show why the internal combustion engine so rapidly became the standard means of propulsion for the horseless carriage, it is necessary to briefly examine the social and economic climate at the end of the 19th century.

The disruptions and horrors of the industrial revolution created a turmoil, changing Europe and North America almost beyond recognition in terms of living and working habits, and to a certain extent the entire social structure. A massive population shift from rural to urban dwelling over a period of eighty years was a major instigator of change. In Britain in 1800, eighty percent of the population lived in the countryside as part of the rural economy. By 1880 a total reversal had taken place with eighty percent of the population living and working in the cities. The economies of the industrial countries of the 1880s bore no relationship at all to those of the previous century. Private capitalism had become the creed and wealth was being created on an unprecedented scale and to some extent shared within society.

This new-found wealth crossed social-class barriers through the rise of industrialists and entrepreneurs, as well as through the success of tradesmen and shopkeepers in the rapidly expanding cities. But the wealthy, or 'carriage trade', were still the minority in relation to the total population, even though their numbers ran into many hundreds of thousands.

By the end of the 19th century the Western world was in the midst, not at the beginning, of an age of technology. All kinds of goods were being produced in massive factories using complex automatic machines. Architecture, bridge-building and tunneling had been revolutionised through the new techniques of producing iron and later steel. In the transport field ships were using mechanical propulsion and steam had found its spiritual home on the railways which now served every continent.

The train was a great leveller, despite distinctions of first, second and third class carriages. The net effect was that anyone could, for a few pence, travel the same route at the same remarkable speed and comfort as the richest in the land. Those who were used to being able to set their own schedules and to move at their own whim, found they had to order their lives by someone else's timetable if they wished to take advantage of the speed and comfort of the trains.

By the last quarter of the 19th century, the only field that remained untouched by the new miracles of science and engineering was that of private door to door transportation. It was, therefore, to this intense social demand from a powerful minority group for private, rather than public transport, that many inventors, entrepreneurs and engineers applied themselves.

What was required was a small light machine, within the financial reach of the privileged few, capable of transporting the individual and family from their door to their destination. A vehicle worthy, in terms of range, speed and comfort, of the astounding technological advances made in other fields.

The pressure must have been immense. After all, people were already overcoming the restrictions of the wheel by taking to the air. The first balloon flights had been achieved more than a hundred years earlier and the technology of powered flight was far advanced. The first practical aeroplanes were built during the first decade of motoring, 1900 to 1910.

It was the internal combustion engine, descended from the gas engine and fueled by petroleum spirit, that provided the essential key, hidden just beyond the grasp of mankind for so long. It was a miracle and it seemed to be almost free. By pouring an apparently small quantity of fuel into a tank, the owner could, by the standards of the day, travel unlimited distances at satisfying speeds.

The only serious obstacle was the poor condition of the roads, but surely with the rapid advances in civil engineering technology they could easily be fixed? After all, hadn't the engineers of the previous hundred years built complexly graded canal systems and railways? Roads suitable for the motor car should present no great problems.

The concept of the internal combustion engine was nothing new. The idea of controlling an explosion within a chamber to provide the force to drive an engine had been the subject of speculation and experimentation for some two hundred years. As early as 1661 Otto von Geuricke had produced sketches of such a machine and in the following decade
Christiaan Huygens described the principles that would be involved in building an internal combustion engine. In 1677 the Abbe Jean de Hautefeuille built an early combustion engine in an attempt to use gunpowder for pumping water. In the 17th century the Frenchman Philippe Lebon patented a coal-gas engine and made the first proposal for electric ignition. That part of his conception came a step closer to reality with the invention of the electric battery by the Italian Alessandro Volta in 1800.

A key date which is ignored in many histories is 1805, the year in which the first vehicle powered by an internal combustion engine moved. The designer, the Swiss Isaac de Rivaz, took out a French patent for his invention two years later.

Then in 1823 Samuel Brown of Britain built an engine with working cylinders, using carburetted hydrogen as fuel, which in 1826 succeeded in climbing Shooters Hill near London. In 1844 Stuart Perry of New York patented a two-cycle internal combustion engine. In 1854 Eugenio Barsanti and Felice Matteucci of Italy took out a British patent for an internal combustion engine and two years later built a practical model of their machine, which was later publicly demonstrated.

What is often described as the 'first successful vehicle powered by an internal combustion engine' was built by a Belgian, Jean-Joseph Etienne Lenoir, in Paris and he obtained a French patent for it the following year. It made numerous journeys between the inventor's Paris home and workshop over a period of many months. In 1862 he fitted his engine to another chassis and in this form he frequently drove between Paris and the Bois de Vincennes. He made a longer excursion of eighteen kilometres in less than one-and-a-half hours. In 1862 Lenoir sold his machine to Czar Alexander II of Russia, achieving the first recorded motor car sale and export with this one act. In the same year a Frenchman, Alphonse Beau de Rochas, invented the four-stroke principle.

In 1864 the gasoline/air two-cycle internal combustion engine powered carriage of Siegfried Marcus came still closer to something recognisable today as a car. It lacked one important though not intrinsic refinement – a clutch. The two rear-driving wheels of the three-wheeler were lifted off the ground by an assistant while the engine was started. As they were lowered the machine got off to a spinning start. An important advance came with the first joint internal combustion engine of Nicholas Otto and Eugene Langen in 1867. It was significant because it was to lead to vehicle manufacture for series sale. In 1872 George B. Brayton took out a U.S. patent for a two-cycle internal combustion engine and in 1876 Nicholas Otto's construction of a four-stroke engine brought the motor car still closer to an industrial reality.

What is often cited as 'the first true petrol internal combustion-engined car' was built in 1883 by Leon-Paul Charles Malandin of France. His vehicle had four wheels and a four-stroke, two cylinder engine.

The machine built by Gottlieb Daimler and Wilhelm Maybach later in the same year has had similar claims made for it. Carl Benz's three-wheeled gas carriage of 1885 is said to have been 'the first vehicle designed and built as a motor car'. Hans Johansen's petrol car, claimed to have been built in the Danish Hammel works around 1886, is significant as the oldest petrol car still running and practical to drive on the road.

Although the car or carriage has been around for several thousands of years, it was the invention and practical application of the internal combustion engine – small, light and powerful enough – that provided, and still provides, the miracle ingredient to convert the horse-drawn carriage into the horseless carriage.

The early theorists and experimenters must certainly be given their due. But, in sum, it is realistic to credit the combined work over a period of twenty years of a number of people including Lenoir, Otto, Maybach, Daimler and Benz with the invention and development of the four-stroke engine. The invention of the motor car as a mechanical device is rarely seen as being as important as its confirmation as a consumer item. Most motoring historians refer to Karl Benz as being 'the father of the motor car' because 'he was the first to sell horseless carriages, made to a set pattern and not one-off experiments, to the public.'

The internal combustion engine made the motor car (and shortly afterwards the aeroplane) and set a standard that no other means of propulsion has yet attained in terms of performance, efficiency and convenience.

The internal combustion engine had made the car feasible, initially for an embryonic market amongst the rich – the 'carriage trade'. But there was also a market for relatively lightweight self-propelled work or commercial vehicles amongst artisans, farmers, travelling salesmen, shopkeepers and in many other areas of commerce, industry and services.

To start with more motor vehicles used for commerce were being sold, but by the 1920s the private car had become numerically vastly greater, so that today less than one in six of all vehicles produced is designated as commercial. Even taking into account all the 'private cars' that are genuinely used as work vehicles, the ratio is still less than one in four.

The early car designers were less concerned with the uses for the intended vehicle than in making horseless carriages that worked reliably and were stable as well as controllable on the road. Between 1885 and 1905 almost every configuration and system for four wheel cars was tried. There has been very little advance except for refinement and re-invention since then.

Before 1905 engines had been placed behind, over and in front of the rear axle. Single, multi-cylinder, vertical, horizontal, V-formed, radial and opposed piston engines had been tried, developed, retained or discarded. Two-strokes, four-strokes, poppet valves, rotary valves, sleeve valves and flap valves had all been tried. Hand steering, foot steering, tiller and lever steering and the steering wheel all had their exponents. Beam axle, De-Dion axles, independent suspension and four-wheel drive, with both in-line and transverse engines all date from before 1905.

The epicyclic gearbox, heart of all modern automatic gearboxes, 2,3,4,5 and 6 speed pinion shaft boxes, variable pulley drive (as used recently in Dafs and some small Volvos), multiple belt and chain transmissions, variable friction drives, fluid couplings all predate 1905. Disk brakes, drum brakes, contracting band brakes, hydraulic controls, rubber, coil, torsion bar and leaf springs; friction, inertia and even hydraulic dampers; solid and pneumatic tyres, cross ply and radial; spark, hot tube, catalyst and compression ignition; separate chassis and monocoque construction; forward control, central and rear driving positions and every possible form of open and closed body had been tried by halfway through the first decade of this century.

The motor car, as we know it, is a 19th century invention to suit 19th century aspirations. It is a mechanical device which obeys the basic laws of statics and dynamics, as true and as well known then as now. The recent use of silicon chips in cars is, so far, strictly superficial and cosmetic; the use of modern plastics simply substitutes oil-based synthetics for metal or wooden parts.

Virtually every basic system and layout used on all modern cars was evolved over eighty years ago. The classic form of car up to the 1970s was established by the French company Panhard & Levassor in 1894. With the engine in front, the gearbox in the middle, rear wheel drive with the passenger compartment situated behind the engine, this is the direct forerunner to the 1982 launched Ford Sierra, successor to the commercially successful and similar Cortina.

It took forty years, until 1934, before another French company, Citroen, established the Traction Avant as a quality quantity-produced model. It is the modern classic form of the car: front engined, front-wheel drive with independent front wheels, a monocoque body/chassis construction with the passenger compartment still situated behind the engine in the middle of the car.

The only other form for the motor car that can claim classic status was actually established in limited production by 1939, about five years after the Traction Avant. It was Dr. Ferdinand Porsche's rear-engined, rear wheel-drive set-up, with the driver and passenger in front of the engine as exemplified in the Volkswagen Beetle and imitated by Renault, Simca, Fiat and Hillman, to mention a few.

As an automobile system the rear-engined car is intrinsically less stable and less predictable, especially in an emergency, than the other two classic forms, although it can be superficially tamed. But it was the cheapest car layout to build. The rear-engined car effectively died with the Corvair debacle at the end of the 1960s, although De Lorean tried, in fairness by mistake, to reintroduce it ten years later.

Porsche's mid-engined layout with rear-wheel drive has its claim to fame in almost every modern racing and sports car as well as many high performance touring cars. It is arguable that for the skilled driver this is the ultimate performance two-wheel drive system if considerations of space, load carrying and accessibility are unimportant.

Within the 19th century concept of the internal combustion engined car there are possibly two classic forms to come. Both offer advantages in stability and safety, as well as efficiency. The four-wheel drive, front-engined motor car is now starting to be introduced at a reasonable price in small quantities. We can still only guess which manufacturer will first make this form universally available.

Beyond the four-wheel drive we will see the three-axled, six-wheel car, so popular with science fiction fantasists. It is an engineering irony that a six-wheeled vehicle can be built lighter and therefore be more efficient and perhaps even cheaper than a four-wheeled vehicle for any given load-carrying requirements. It is no coincidence that six-wheeled Grand Prix cars appear from time to time and a number of electric vehicles use an extra axle to support the weight of the batteries more effectively. The potential advantages in adhesion, ride, braking and load-carrying are so great that six-wheelers could become the first truly radical change in the motor car since the tradition was forged in the 1880s and 90s.

While most of the early car-makers were busying themselves with the mechanical systems and layouts, there were a few who had a distinctly philosophical as well as technical approach, designers who sought to produce a car for the people who would use it.

Outstanding amongst these was Dr Fredrick Lanchester, an Englishman born in 1868. Lanchester was trained as an engineer and made the aerodynamics of flight one of his special subjects, some twenty years before the advent of powered aircraft.

Lanchester designed and produced the first wholly English car in 1896. This led to a production model in 1900, a vehicle which was more like a rolling test bed than a production car, and yet in it were combined a series of rational ideas that any car designer today would do well to study.

Lanchester's vehicle was revolutionary. He designed the suspension to have a natural frequency, similar to the rise and fall of walking, a frequency that the human body was used to. He had, therefore, soft springs with large vertical wheel movements. His car was practical, with the weight concentrated at the centre, high ground clearance but low centre of gravity, due to the horizontal mounting of the engine. The driver's eye line was set close to that of a walking person so the road conditions ahead would be interpreted from a familiar position.

Lanchester used a semi-monocoque chassis/body construction, now understood to be the lightest and strongest way to build a road vehicle. He placed the engine in the middle of the car, a position that is theoretically ideal for a rear-wheel driven vehicle. He did much pioneering work on the balancing of engines and this car probably had the first fully-balanced vibration-free engine in the world – one of the few ever built. He applied the principles of epicyclic gear chain to the gearbox and invented and used the disk brake. He also used a worm-drive rear axle, roller bearings, an improved carburettor and other original items.

Among his many achievements, Lanchester was the first to design a car with absolute interchangeability of precision mechanical components, an essential step on the road to mass production. Commercially, however, he failed, losing his position in the company that bore his name and becoming, in the words of one commentator (L.J.K. Setright in The Designers) 'the most accomplished gentleman ever wasted on the motor industry.'

His huge contribution to the concept of the motor vehicle was largely ignored except in his detailed engineering solutions. Many of his ideas, like the disk brake and his suspension philosophy, had to wait decades before being re-invented. So what went wrong?

Lanchester insisted on his own ergonomically designed control system, which involved tiller steering. He dogmatically stuck to this steering method until 1910, long after it had been accepted that the steering wheel was not only a better system, but was essential to control cars at the speeds they were attaining. Lanchester fought with his backers, resigning in 1909 to hand over to his brother George. He, in turn, was to fail in the Depression, having opted for the heavy, luxury carriage market.

Lanchester typified a certain breed of car maker, the individual designer/engineer. As reluctant entrepreneurs they have had an unenviable record of commercial failure and bankruptcy. This is probably to be expected but unfortunately they also often have a poor showing in the field of improving the car. Their failures have been used time and again as absolute proof that some idea or other will not work technically or in the market place. It could be that the pressures of dealing with demanding financial backers made many of the car industry's idealists, dogmatic and determined to hang themselves on a string of fixed ideas. This seems to have been the case with Lanchester. In any event, it was left to others, with a better mix of technical and entrepreneurial skills, to introduce most of the engineering developments in commercially successful vehicles.

THE EARLY DAYS OF POWERED TRANSPORT

3000 BC	Earliest known sailing boats in Egypt
2000 BC	Pharaoh Amenhemet III drives a landsailer
1st Century AD	Hero of Alexandria designs a steam aeolipile carriage
1420	Giovanni Fontana designs a hand-cranked locomotive
1475	Robert Valturio designs a vehicle to be driven by a windmill through gears to the wheels
1479	Gilles de Dom receives 25 livres from the City of Antwerp for a mechanically propelled (not powered) vehicle
14..	Leonardo da Vinci designs a clockwork car
1599	Simon Stevin runs large wind carriages on the beaches of the Netherlands
1629	Giovanni Branca builds a steam carriage
16..	Otto von Guericke builds an air pump, comprising metal pistons, cylinders and connecting rods

1662	Ferdinand Verbiest builds a steam engine, suggestive of the modern steam turbine
1677	Abbe Jean de Hautefeuille attempts to pump water using a machine powered by the combustion of gunpowder
1680	Isaac Newton designs a wagon powered by steam propulsion
1690	Denis Papin designs an atmospheric steam engine
1696	Thomas Savery builds the first functioning steam pump
1698	Denis Papin builds a steam-powered carriage
1705	Thomas Newcomen invents the beam steam engine
1707	Denis Papin builds a steam-powered boat
1740	Jacques de Vauconson runs a clockwork carriage on the streets of Paris
1769	Nicholas-Joseph Cugnot builds 'the first true automobile', a steam-powered vehicle
1769	James Watt designs an improved steam pumping engine
1776	Claude Francois, the Marquis de Jouffroy d'Abbans, installs a single-acting steam pump in an experimental steamboat

LATER DEVELOPMENTS IN STEAM POWER

1690	Denis Papin invents an atmospheric steam pump
1698	Thomas Savery builds the first steam pump
1698	Denis Papin's steam carriage
1707	Denis Papin runs a successful experimental steam boat
1712	Thomas Newcomen invents the beam steam engine
1769	James Watt's improved steam pumping engine
1776	Claude Francois, the Marquis de Jouffroy d'Abbans, installs a single-acting steam engine in a boat, which runs experimentally
1782	James Watt invents the double-acting steam engine
1783	Claude Francois, the Marquis de Jouffroy d'Abbans' improved steamboat Pyroscaphe makes a successful journey
1784	William Murdock builds and runs a model steamboat on the Delaware River
1787	Oliver Evans obtains a patent to run a steam vehicle on the public roads of Maryland
1788	Robert Fourness demonstrates a model, three-cylinder steam tractor
1790	Charles Dallery builds a successful steam carriage
1792	Oliver Evans invents a high-powered steam engine
1798	Richard Trevithick's high pressure, non-condensing engine used in a carriage
1802	The Charles Dundas steamer plies regularly on the Firth and Clyde canal
1804	Richard Trevithick first uses a steam vehicle on rails
1805	Oliver Evans fits his steam-powered dredger Orukter Amphibolos with wheels and runs it on land
1821	The first iron steamship the Aaron Manby built
1825	Opening of the Stockton and Darlington Railway, the first steam-powered common carrier
1829	Sir Goldsworthy Gurney inaugurates a long distance steam coach service
1830	Opening of the first long distance railway, Liverpool to Manchester
1838	The Sirius makes the first mainly steam-powered Atlantic crossing
1839	Launching of the Robert F. Stockton, the first propeller-driven steamship

HISTORY OF THE COMBUSTION ENGINE

1661	Otto von Guericke sketches an internal combustion engine
1677	Christiaan Huygens describes the principles of internal combustion
1677	Jean de Hautefeuille makes an unsuccessful attempt to pump water by a controlled explosion of gunpowder
1791	John Barber describes an air and paraffin gas turbine
1794	Robert Street registers a British patent describing an internal combustion engine using pistons and levers
1801	Philippe Lebon proposes compressing the charge and electric ignition
1805	Isaac de Rival builds a working model of a gas-powered internal combustion engine, propelling a toy cart

1824	Sadi Carnot outlines the fundamental theory of the internal combustion engine
1826	Samuel Brown's 'gas and vacuum' engine, powered by two cylinders linked by a rocking beam, drives a vehicle up Shooters Hill
1844	Stuart Perry patents a two–cycle internal combustion engine in the U.S.
1854	Eugenio Barsanti and Felice Matteucci obtain a British patent for a coal–gas engine
1859	Eugenio Barsanti and Felice Matteucci test run a model internal combustion engine
1860	Jean–Joseph Etienne Lenoir builds a vehicle powered by an internal combustion engine with working cylinders
1862	Alphonse Beau de Rochas describes the operating cycle of the four–stroke internal combustion engine
1864	Siegfried Marcus builds a two–cycle petrol internal combustion engine
1866	Nikolaus Otto and Eugene Langen build and patent an industrial free–piston internal combustion engine
1876	Nikolaus Otto and Eugene Langen build a four–stroke, 2–cylinder petrol engine to run a vehicle
1883	Levin Paul Charles Malandin builds a four–stroke, 2–cylinder gasoline internal combustion engine to run a car
1883	Gottleib Daimler and Wilhelm Maybach build a high–speed internal combustion engine fuelled by petrol
1885	Karl Benz builds a vehicle powered by a petrol internal combustion engine, with a light tubing frame, designed as a car
1886	Hans Johansen builds an internal combustion engine to drive a motor car, which is still in running order
1901	Alberto Santos Dumont makes a successful petrol engine–powered balloon flight

THE TECHNICAL DEVELOPMENT OF THE MODERN CAR

1894	Panhard et Levassor – first classic car, engine at the front, gearbox behind the engine, rear wheel drive
1898	Louis Renault – the first propeller shaft driven car, direct drive top gear
1900	Lanchester – first fully balanced engine, disc brake, epicyclic gearbox, leading and trailing low rate suspension, interchangeable precision parts
1907	Model T Ford – multi–purpose vehicle with articulated suspension, epicyclic gearbox, light weight alloy steel construction. First real mass production vehicle
1920	Half the vehicles in the world are Model T Fords
1934	Citroen Traction Avant – First mass–produced monocoque, front–wheel drive, independent front suspension – trailing arm rear suspension, torsion bars, hydraulic brakes
1935/36	Citroen 2CV prototype – interconnected rising rate independent suspension, innertia dampers, front–wheel drive, air cooled, multi–purpose light weight body
1954	DS Citroen – active self levelling, rising rate fully independent hydropneumatic suspension, self levelling lights, clutchless gear change, removable composite body – fibreglass, aluminium, stainless steel, plastic and steel. CD factor – approx .33, inboard disc brakes, automatic variable braking between front and rear to compensate for load, zero scrub power steering, positive power braking, energy absorbing crash protection sections
1960	Jensen FF – Dunlop 'Maxaret' anti lock braking – Ferguson type permanent four–wheel drive transmission

The basic AFRICAR specification has changed very little from the original design concept which I laid down in 1979/1980. The track, the ground clearance, the weight, the plywood epoxy chassis and body panels, leading and trailing suspension links, the type of spring and kind of sub frame structures were all there from the beginning. The engine went through a number of changes. It started as a Boxer–type horizontally opposed four–stroke with rotary valves which were later changed to poppet valves. In Fiji in 1985, I finally opted for the opposed–piston, supercharged two–stroke format, a type of engine frequently used over the years for everything from light and heavy trucks to boats and aircraft.

The transmission was initially based on an infinitel

variable belt drive, and in this form it is still retained in the rear differential system. A more conventional main gearbox has been developed to handle the power of the bigger engines and the six-wheelers, but work continues on a dry belt box mounted directly behind the engine.

In the development of AFRICAR I have been helped by a number of people. My own engineering experience stopped at the beginning of the 1960s when my career turned to photography and film making. So at the end of the 1970s I needed a crash course to 'de-rust' me. Professional consultants who contributed most to AFRICAR are Tony Best for the suspension, Bill Bonner for translating the original engine drawings into reality and Bill Bonner again for the main transaxle gearbox. Rod Quaife for the steering rack, and in the first prototype stage Valerian Dare-Bryan for prototype suspension arms, beams, subframes and a lot of the original experimental construction.

The very first plywood chassis was constructed in the workshops of cabinet-maker Richard Beecher, in London, and the first complete vehicle was put together at Nick Oroussoff's prototyping workshop in Kew.

Despite this essential specialist input there is one person without whom I could not have so freely developed my ideas and seen them come to reality and then changed my mind and tried again. John Fitzpatrick, now Development Manager of AFRICAR, must bear a lot of the blame for the effective way in which the vehicle works. It is significant and possibly unique that the three prototype vehicles, while on the Arctic to Equator journey, did not suffer a single fracture at a weld or in any of the fabrication personally made by John.

I have to take full responsibility for the plywood and resin chassis which, dare I say it, also gave us no problems in Africa.

AFRICAR SPECIFICATION

ENGINE
Type: Twin crankshaft opposed piston supercharged two-stroke with through-flow scavenging characteristics
Fuel: Diesel or petrol
Cylinders: Two or three
Cubic capacity: 1.3 litres or 2 litres
Bore and stroke: 75mm x 75mm
Compression ratio: Petrol 9.5 to 1
Diesel 16 to 1
Cooling: Air cooled
Lubrication: Dry sump with twin scavenging pumps and oil cooler with separate oil reservoir
Supercharger: Rootes-type

TRANSMISSION
Gearbox: Manual with transfer box. Primary gearbox four speed. Transfer box two or three speed. Permanent four-wheel drive
Front Differential: Standard type housed in gearbox casing, option of semi-locking suitable for front-wheel drive
Rear Differential: Variable speed pulley drive to each rear wheel allowing both differential effect and freedom from transmission wind-up
Drive Shafts: Plunging spline
C.V. Joints: Birfield Rezeppa inboard and outboard
Clutch: Single plate, cable operated. Air cooled
Steering: Rack and pinion
Hub Geometry: Zero scrub radius
Steering Wheel Turns: Lock to lock 3.5
Turning Circle: 36 feet

BRAKES
Front: Inboard discs
Rear: Inboard discs/drums
Parking Brake: Transmission disc/drum

SUSPENSION
Front: Leading arms
Rear: Trailing arms
Characteristics: Low rate, heavily damped with rising rate to compensate for load. Interconnected front to rear

Springs: Nitrogen gas spheres
Dampers: Hydraulic integral with springs

WHEELS & TYRES
Wheels: 15 inch diameter, 5 inch rims with either three or six studs
Tyres: 195 x 15 steel belted radials M S

ELECTRICAL EQUIPMENT
Generator: Gear driven alternator
Lights: Two main headlights and two spotlights.
Battery: Two heavy-duty sealed units
Alternative: Starting handle

FUEL
Capacity: 112 litres or approx 25 gallons
Maximum Range: 750 miles – plus

CHASSIS
Chassis: Plywood, glued, filleted and sealed with epoxy resin.
Sub frames: Zinc coated steel
Passenger Protection: Steel roll cage throughout passenger compartment. Energy absorbing bumpers and sub frames Foam filled chassis side members
Body Panels: Plywood sealed with epoxy
Body Forms: Station wagon; mini bus; van; ambulance; high top van; full width van; single cab pick-up; full cab pick-up; flat bed and platforms.

DIMENSIONS AND PERFORMANCE
Track: 1600 mm
Wheel base: 2500 mm
Ground Clearance: 310 mm
Width: 1800 mm
Length and Height: Dependent on model
Seating: Three to six depending on model, up to 17 in the case of mini buses

LOAD CAPACITY
Four-Wheelers: One tonne
Six-wheelers Two tonnes
Cruising speeds: 75 mph to 95 mph depending on model and engine.

AN 0533071 8